P9-CAD-569

CYBER

WAR

CYBER WAR

THE NEXT THREAT
TO NATIONAL SECURITY AND
WHAT TO DO ABOUT IT

RICHARD A. CLARKE
AND ROBERT K. KNAKE

An Imprint of HarperCollinsPublishers

HarperCollins books may be purchased for educational, business, or sales promotional use. For information, please write: Special Markets Department, HarperCollins Publishers, 10 East 53rd Street, New York, NY 10022.

FIRST EDITION

Designed by Mary Austin Speaker

Library of Congress Cataloging-in-Publication Data has been applied for.

ISBN: 978-0-06-196223-3

10 11 12 13 14   ov/wcf   10 9 8 7 6 5 4 3 2 1

To the late William Weed Kaufmann, who taught me and so many others how to analyze national security issues.

RICHARD CLARKE

To my wife, Elizabeth, whose support knows no bounds. And for our daughter, Charlotte, may you grow up in a more peaceful century.

ROB KNAKE

CONTENTS

Introduction

It was in the depths of a gray and chill Washington winter. On a side street not far from Dupont Circle, in a brownstone filled with electric guitars and an eclectic collection of art, we gathered to remember the man who had taught us how to analyze issues of war and defense. Two dozen of his former students, now mostly in their fifties, drank toasts that February night in 2009 to Professor William W. Kaufmann, who had died weeks earlier at age ninety. Bill, as everyone referred to him that night, had taught defense analysis and strategic nuclear weapons policy at MIT for decades, and later at Harvard and the Brookings Institution. Generations of civilian and military "experts" had earned that title by passing through his courses. Bill was also an advisor to six Secretaries of Defense, sitting in the "front office" on the E Ring of the Pentagon. He shuttled between Boston and Washington every week for decades.

Behind his back, some of us had referred to Professor Kaufmann as "Yoda," in part because of a vague physical and stylistic resemblance, but chiefly because we thought of him as our Jedi master, the man who understood the workings of the Force and tried to teach them to us. As an analyst and advisor, Bill had been one of a handful of civilians who had created the framework of strategic nuclear war doctrine in the late 1950s and early 1960s. They had walked the United States back from a nuclear strategy that had called for the United States to go first in a nuclear war, to use all of its nuclear weapons in one massive attack, and to destroy hundreds of cities in Europe and Asia. Bill and his colleagues had probably prevented a global nuclear war and had made strategic arms control possible. Our conversation that night, lubricated by the same martinis Bill used to drink with us, turned to the future. What could we do to honor the memory of William W. Kaufmann and the other strategists of the second half of the twentieth century? We could, someone suggested, continue their work, use what Bill had taught us, ask the tough analytical questions about today's strategy. Another at the table suggested that today is very different from the 1950s, when nuclear weapons were being deployed without a thoughtful strategy; strategies are well developed today.

But is it such a different time? In the first decade of the twenty-first century, the U.S. developed and systematically deployed a new type of weapon, based on our new technologies, and we did so without a thoughtful strategy. We created a new military command to conduct a new kind of high-tech war, without public debate, media discussion, serious congressional oversight, academic analysis, or international dialogue. Perhaps, then, we are at a time with some striking similarities to the 1950s. Perhaps, then, we need to stimulate learned discussion and rigorous analysis about that new kind of weapon, that new kind of war.

It is cyberspace and war in it about which I speak. On October 1, 2009, a general took charge of the new U.S. Cyber Command, a military organization with the mission to use information technology and the Internet as a weapon. Similar commands exist in Russia, China, and a score of other nations. These military and intelligence organizations are preparing the cyber battlefield with things called "logic bombs" and "trapdoors," placing virtual explosives in other countries in peacetime. Given the unique nature of cyber war, there may be incentives to go first. The most likely targets are civilian in nature. The speed at which thousands of targets can be hit, almost anywhere in the world, brings with it the prospect of highly volatile crises. The force that prevented nuclear war, deterrence, does not work well in cyber war. The entire phenomenon of cyber war is shrouded in such government secrecy that it makes the Cold War look like a time of openness and transparency. The biggest secret in the world about cyber war may be that at the very same time the U.S. prepares for offensive cyber war, it is continuing policies that make it impossible to defend the nation effectively from cyber attack.

A nation that has invented the new technology, and the tactics to use it, may not be the victor, if its own military is mired in the ways of the past, overcome by inertia, overconfident in the weapons they have grown to love and consider supreme. The originator of the new offensive weaponry may be the loser unless it has also figured out how to defend against the weapon it has shown to the rest of the world. Thus, even though the American colonel Billy Mitchell was the first to understand the ability of small aircraft to sink mighty battleships, it was the Japanese Imperial Navy that acted on that understanding, and came close to defeating the Americans in the Pacific in World War II. It was Britain that first developed the tank, and a French colonel, Charles de Gaulle, who devised the tactics of rapid attack with massed tanks, supported by air and artillery. Yet it was a recently defeated Germany that perfected the tank in the 1930s and first employed de Gaulle's tactics, which later became known as blitzkrieg. (As recently as 1990, and again in 2003, the

U.S. military went to war with an updated version of the seventy-year-old blitzkrieg tactic: fast movement of heavy tank units, supported by aircraft.)

Warmed by the camaraderie of my fellow ex-students, and by the martinis, I left the brownstone and wandered out into that cold night, pondering this irony of history, and making a commitment to myself, and to Bill, that I would try to stimulate open, public analysis and discussion of cyber-war strategy before we stumbled into such a conflict. This book is the down payment on that commitment. I knew that I needed a younger partner to join me in trying to understand the military and technological implications of cyber war well enough to produce this book. Different generations think of cyberspace differently. For me, looking at my sixtieth birthday in 2010, cyberspace is something that I saw gradually creep up around me. It happened after I had already had a career dealing with nuclear weapons, in a bipolar world. I became the first Special Advisor to the President for Cyber Security in 2001, but my views of cyber war are colored by my background in nuclear strategy and espionage.

Rob Knake was thirty when he and I wrote this book. For his generation, the Internet and cyberspace are as natural as air and water. Rob's career has focused on homeland security and the transnational threats of the twenty-first century. We have worked together at Harvard's Kennedy School of Government, at Good Harbor Consulting, and on the Obama for America campaign. In 2009, Rob won the prestigious International Affairs Fellowship at the Council on Foreign Relations with an appointment to study cyber war. We decided to use the first-person singular in the text because many times I will be discussing my personal experiences with government, with the information-technology industry, and with Washington's clans, but the research, writing, and concept development were a joint enterprise. We have wandered around Washington and other parts of this country together in search of answers to the many questions surrounding cyber war. Many people have helped us in that search, some of them wishing to remain unnamed in this book because of their past or present associations. We had spent long hours discussing, debating, and arguing until we found a synthesis of our views. Rob and I both agree that cyber war is not some victimless, clean, new kind of war that we should embrace. Nor is it some kind of secret weapon that we need to keep hidden from the daylight and from the public. For it is the public, the civilian population of the United States and the publicly owned corporations that run our key national systems, that are likely to suffer in a cyber war.

While it may appear to give America some sort of advantage, in fact cyber war places this country at greater jeopardy than it does any other nation. Nor is this new kind of war a game or a figment of our imaginations. Far from being an alternative to conventional war, cyber war may actually increase the likelihood of the more traditional combat with explosives, bullets, and missiles. If we could put this genie back in the bottle, we should, but we can't. Therefore, we need to embark on a complex series of tasks: to understand what cyber war is, to learn how and why it works, to analyze its risks, to prepare for it, and to think about how to control it.

This book is an attempt to begin to do some of that. It is not a technical book, not meant to be an electrical engineer's guide to the details of cyber weapons. Nor is it designed to be a Washington wonk's acronym-filled, jargon-encrusted political or legal exegesis. Finally, it is also definitely not a military document and not written to be immediately translatable into Pentagonese. Therefore, some experts in each of those fields may think the book simplistic in places where it discusses things they understand and opaque in parts that stretch beyond their expertise. Overall, we have tried to strike a balance and to write in an informal style that will be both clear and occasionally entertaining. Lest you take too much comfort in those assurances, however, it is necessary in a book on this subject to discuss the technology, the ways of Washington, as well as some military and intelligence themes. Likewise, it is impossible to avoid entirely the use of acronyms and jargon, and therefore we include a glossary (starting on page 281).

I have been taught by senior national security officials for decades never to bring them a problem without also suggesting a solution. This book certainly reveals some problems, but it also discusses potential solutions. Putting those or other defenses in place will take time, and until they are a reality, this nation and others are running some new and serious risks to peace, to international stability, to internal order, and to our national and individual economic well-being.

The authors wish to thank the many people who helped us with this book, most important the experts in and out of governments who helped us on condition that they go unnamed. Pieter Zatko, John Mallery, Chris Jordan, Ed Amoroso, Sami Saydjari, and Barnaby Page helped us understand some of the more technical aspects of cyber security. Paul Kurtz served as a constant sounding board and helped shape our thinking in innumerable ways. Ken Minihan, Mike McConnell, and Rich Wilhelm gave us added insight from their decades in government and the private sector, Alan Paller, Greg Rattray, and Jim Lewis gave their insights and latest thinking on this complex topic. We thank Janet Napolitano for taking time out of her busy schedule to meet with us and for being willing to do so on the record. We also thank Rand Beers for his wisdom. Will Howerton helped in a major way to get this book across the finish line. He possesses a keen editorial eye and a gift for research. Will Bardenwerper also provided editorial assistance.

Bev Roundtree, as she has been on so many projects over the decades, was the sine qua non.

CHAPTER ONE

TRIAL RUNS

A quarter-moon reflected on the slowly flowing Euphrates, a river along which nations have warred for five thousand years. It was just after midnight, September 6, 2007, and a new kind of attack was about to happen along the Euphrates, one that had begun in cyberspace. On the east side of the river, seventy-five miles south into Syria from the Turkish border, up a dry wadi from the riverbank, a few low lights cast shadows on the wadi's sandy walls. The shadows were from a large building under construction. Many North Korean workers had left the construction site six hours earlier, queuing in orderly lines to load onto buses for the drive to their nearby dormitory. For a construction site, the area was unusually dark and unprotected, almost as if the builder wanted to avoid attracting attention.

Without warning, what seemed like small stars burst above the site, illuminating the area with a blue-white clarity brighter than daylight. In less than a minute, although it seemed longer to the few Syrians and Koreans still on the site, there was a blinding flash, then a concussive sound wave, and then falling pieces of debris. If their hearing had not been temporarily destroyed by the explosions, those on the ground nearby would then have heard a longer acoustic wash of military jet engines blanketing the area. Had they been able to look beyond the flames that were now sweeping the construction site, or above the illuminating flares that were still floating down on small parachutes, the Syrians and Koreans might have seen F-15 Eagles and F-16 Falcons banking north, back toward Turkey. Perhaps they would even have made out muted blue-and-white Star of David emblems on the wings of the Israeli Air Force strike formation as it headed home, unscathed, leaving years of secret work near the wadi totally destroyed.

Almost as unusual as the raid itself was the political silence that followed. The public affairs offices of the Israeli government said nothing. Even more telling, Syria, which had been bombed, was silent. Slowly, the story started to emerge in American and British media. Israel had bombed a complex in eastern Syria, a facility being built by North Koreans. The facility was related to weapons of mass destruction, the news accounts reported from unnamed sources. Israeli press censors allowed their nation's newspapers to quote American media accounts, but prohibited them from doing any reporting of their own. It was, they said, a national security matter. Prompted by the media accounts, the Syrian government belatedly admitted there had been an attack on their territory. Then they protested it, somewhat meekly. Syrian President Assad asserted that what had been destroyed was "an empty building." Curiously, only North Korea joined Damascus in expressing outrage at this surprise attack.

Media accounts differed slightly as to what had happened and why, but most quoted Israeli government sources as saying that the facility had been a North Korean–designed nuclear weapons plant. If that was true, North Korea had violated an agreement with the United States and other major powers that it would stop selling nuclear weapons know-how. Worse, it meant that Syria, a nation on Israel's border, a nation that had been negotiating with Israel through the Turks, had actually been trying secretly to acquire nuclear weapons, something that even Saddam Hussein had stopped doing years before the U.S. invasion of Iraq.

Soon, however, self-anointed experts were casting doubt on the "Syria was making a nuclear bomb" story.

Satellite pictures, taken by reconnaissance satellite, were revealed by Western media. Experts noted that the site had little security around it before the bombing. Some contended that the building was not tall enough to house a North Korean nuclear reactor. Others pointed to the lack of any other nuclear infrastructure in Syria. They offered alternative theories. Maybe the building was related to Syria's missile program. Maybe Israel had just gotten it wrong and the building was relatively innocent, like Saddam Hussein's alleged "baby milk factory" of 1990 or Sudan's supposed aspirin plant of 1998, both destroyed in U.S. strikes. Or maybe, said some commentators, Syria was not the real target. Maybe Israel was sending a message to Iran, a message that the Jewish state could still successfully carry out surprise air strikes, a message that a similar strike could occur on Iranian nuclear facilities unless Tehran stopped its nuclear development program.

Media reports quoting unnamed sources claimed various degrees of American involvement in the raid: the Americans had discovered the site on satellite photography, or the Americans had overlooked the site and the Israelis had found it on satellite images given to them routinely by the U.S. intelligence community; the Americans had helped plan the bombing, perhaps persuading the Turkish military to look the other way as the Israeli attack formation sailed over

Turkey to surprise Syria by attacking from the north. Americans—
or were they Israelis?—had perhaps snuck into the construction
site before the bombing to confirm the North Korean presence,
and maybe verify the nuclear nature of the site. President George
W. Bush, uncharacteristically taciturn, flatly refused to answer a
reporter's question about the Israeli attack.

The one thing that most analysts agreed upon was that something
strange had happened. In April 2008, the CIA took the unusual
step of producing and publicly releasing a video showing clandestine
imagery from inside the facility before it was bombed. The film left
little doubt that the site had been a North Korean–designed nuclear
facility. The story soon faded. Scant attention was paid when, seven
months later, the UN's International Atomic Energy Agency (IAEA)
issued its report. It had sent inspectors to the site. What the inspectors found was not a bombed-out ruin, nor did they come upon a
beehive of renewed construction activity. Instead, the international
experts were taken to a site that had been neatly plowed and raked,
a site showing no signs of debris or construction materials. It looked
like an unimproved home lot for sale in some desert community
outside of Phoenix, perfectly anodyne. The disappointed inspectors
took pictures. They filled plastic ziplock baggies with soil samples
and then they left the banks of the Euphrates and flew back to their
headquarters on an island in the Danube near Vienna. There they
ran tests in their laboratories.

The IAEA announced, again to little attention, that the soil samples had contained unusual, "man-made," radioactive materials. For
those few who had been following the mystery of Syria's Euphrates
enigma, that was the end of the story, vindicating Israel's highly
regarded intelligence service. Despite how unlikely it seemed, Syria
in fact had been secretly fooling around with nuclear weapons, and
the bizarre regime in North Korea had been helping. It was time to
reassess the intentions of both Damascus and Pyongyang.

Behind all of this mystery, however, was another intrigue. Syria had spent billions of dollars on air defense systems. That September night, Syrian military personnel were closely watching their radars. Unexpectedly, Israel had put its troops on the Golan Heights on full alert earlier in the day. From their emplacements on the occupied Syrian territory, Israel's Golani Brigade could literally look into downtown Damascus through their long-range lenses. Syrian forces were expecting trouble. Yet nothing unusual appeared on their screens. The skies over Syria seemed safe and largely empty as midnight rolled around. In fact, however, formations of Eagles and Falcons had penetrated Syrian airspace from Turkey. Those aircraft, designed and first built in the 1970s, were far from stealthy. Their steel and titanium airframes, their sharp edges and corners, the bombs and missiles hanging on their wings, should have lit up the Syrian radars like the Christmas tree illuminating New York's Rockefeller Plaza in December. But they didn't.

What the Syrians slowly, reluctantly, and painfully concluded the next morning was that Israel had "owned" Damascus's pricey air defense network the night before. What appeared on the radar screens was what the Israeli Air Force had put there, an image of nothing. The view seen by the Syrians bore no relation to the reality that their eastern skies had become an Israeli Air Force bombing range. Syrian air defense missiles could not have been fired because there had been no targets in the system for them to seek out. Syrian air defense fighters could not have scrambled, had they been fool enough to do so again against the Israelis, because their Russian-built systems required them to be vectored toward the target aircraft by ground-based controllers. The Syrian ground-based controllers had seen no targets.

By that afternoon, the phones were ringing in the Russian Defense Ministry off Red Square. How could the Russian air defense system have been blinded? Syria wanted to know. Moscow promised to send experts and technicians right away. Maybe there had been an implementation problem, maybe a user error, but it would be fixed immediately. The Russian military-industrial complex did not need that kind of bad publicity about its products. After all, Iran was about to buy a modern air defense radar and missile system from Moscow. In both Tehran and Damascus, air defense commanders were in shock.

Cyber warriors around the world, however, were not surprised. This was how war would be fought in the information age, this was Cyber War. When the term "cyber war" is used in this book, it refers to actions by a nation-state to penetrate another nation's computers or networks for the purposes of causing damage or disruption. When the Israelis attacked Syria, they used light and electric pulses, not to cut like a laser or stun like a taser, but to transmit 1's and 0's to control what the Syrian air defense radars saw. Instead of blowing up air defense radars and giving up the element of surprise before hitting the main targets, in the age of cyber war, the Israelis ensured that the enemy could not even raise its defenses.

The Israelis had planned and executed their cyber assault flawlessly. Just how they did it is a matter of some conjecture.

There are at least three possibilities for how they "owned" the Syrians. First, there is the possibility suggested by some media reports that the Israeli attack was preceded by a stealthy unmanned aerial vehicle (UAV) that intentionally flew into a Syrian air defense radar's beam. Radar still works essentially the same way it began seventy years ago in the Battle of Britain. A radar system sends out a directional radio beam. If the beam hits anything, it bounces back to a receiver. The processor then computes where the object was that the radio beam hit, at what altitude it was flying, at what speed it was moving, and maybe even how big an object was up there. The key fact here is that the radar is allowing an electronic beam to come from the air, back into the ground-based computer system.

Radar is inherently an open computer door, open so that it can receive back the electronic searchers it has sent out to look for things in the sky. A stealthy Israeli UAV might not have been seen by the Syrian air defense because the drone would have been coated with material that absorbs or deflects a radar beam. The UAV might, however, have been able to detect the radar beam coming up from the ground toward it and used that very same radio frequency to transmit computer packets back down into the radar's computer and from there into the Syrian air defense network. Those packets made the system malfunction, but they also told it not to act there was anything wrong with it. They may have just replayed a do-loop of the sky as it was before the attack. Thus, while the radar beam might later have bounced off the attacking Eagles and Falcons, the return signal did not register on the Syrian air defense computers. The sky would look just like it had when it was empty, even though it was, in actuality, filled with Israeli fighters. U.S. media reports indicate that the United States has a similar cyber attack system, code-named Senior Suter.

Second, there is the possibility that the Russian computer code controlling the Syrian air defense network had been compromised by Israeli agents. At some point, perhaps in the Russian computer lab or in a Syrian military facility, someone working for Israel or one of its allies may have slipped a "trapdoor" into the millions of lines of computer code that run the air defense program. A "trapdoor" (or "Trojan Horse") is simply a handful of lines of computer code that look just like all the other gibberish that comprise the instructions for an operating system or application. (Tests run by the National Security Agency determined that even the best-trained experts could not, by visually looking through the millions of lines of symbols, find the "errors" that had been introduced into a piece of software.)

The "trapdoor" could be instructions on how to respond to certain circumstances. For example, if the radar processor discovers a particular electronic signal, it would respond by showing no targets in the sky for a designated period of time, say, the next three hours. All the Israeli UAV would have to do is send down that small electronic signal. The "trapdoor" might be a secret electronic access point that would allow someone tapping into the air defense network to get past the intrusion-detection system and firewall, through the encryption, and take control of the network with full administrator's rights and privileges.

The third possibility is that an Israeli agent would find any fiber-optic cable of the air defense network somewhere in Syria and splice into the line (harder than it sounds, but doable). Once on line, the Israeli agent would type in a command that would cause the "trapdoor" to open for him. While it is risky for an Israeli agent to be wandering around Syria cutting into fiber-optic cables, it is far from impossible. Reports have suggested for decades that Israel places its spies behind Syrian borders. The fiber-optic cables for the Syrian national air defense network run all over the country, not just inside military installations. The advantage of an agent in place hacking into the network is that it does not cause the operation to rely upon the success of a "takeover packet" entering the network from a UAV flying overhead. Indeed, an agent in place could theoretically set up a link from his location back to Israel's Air Force command post. Using low-probability-of-intercept (LPI) communications methods, an Israeli agent may be able to establish "cove comms" (covert communications), even in downtown Damascus, beaming up to a satellite with little risk of anyone in Syria noticing.

Whatever method the Israelis used to trick the Syrian air defense network, it was probably taken from a playbook they borrowed from the U.S. Our Israeli friends have learned a thing or two from the programs we have been working on for more than two decades. In 1990, as the United States was preparing to go to war with Iraq for the first time, early U.S. cyber warriors got together with Special Operations commandos to figure out how they could take out the extensive Iraqi air defense radar and missile network just before the initial waves of U.S. and allied aircraft came screeching in toward Baghdad. As the hero of Desert Storm, four-star General Norm Schwarzkopf, explained to me at the time, "these snake-eaters had some crazy idea" to sneak into Iraq before the first shots were fired and seize control of a radar base in the south of the country. They planned to bring with them some hackers, probably from the U.S. Air Force, who would hook up to the Iraqi network from inside the base and then send out a program that would have caused all the computers on the network all over the country to crash and be unable to reboot.

Schwarzkopf thought the plan risky and unreliable. He had a low opinion of U.S. Special Operations Command and feared that the commandos would become the first Americans held as prisoners of war, even before the war started. Even worse, he feared that the Iraqis would be able to turn their computers back on and would start shooting down some of the two thousand sorties of attacks he planned for the first day of the air war. "If you want to make sure their air defense radars and missiles don't work, blow them up first. That way they stay dead. Then go in and bomb your targets." Thus, most of the initial U.S. and allied air sorties were not bombing raids on Baghdad headquarters or Iraqi Army divisions, they were on the air defense radar and missile sites. Some U.S. aircraft were destroyed in those attempts, some pilots were killed, and some were taken prisoner.

When, thirteen years later, the U.S. went to war with Iraq a second time, well before the initial waves of American fighter-bombers swept in, the Iraqi military knew that their "closed-loop" private, secure military network had already been compromised. The Americans told them.

Thousands of Iraqi military officers received e-mails on the Iraqi Defense Ministry e-mail system just before the war started.

Although the exact text has never been made public, several reliable sources revealed enough of the gist to reconstruct what you might have read had you been, say, an Iraqi Army brigadier general in charge of an armored unit outside of Basra. It would have read something like this:

> This is a message from United States Central Command. As you know, we may be instructed to invade Iraq in the near future. If we do so, we will overwhelm forces that oppose us, as we did several years ago. We do not want to harm you or your troops. Our goal would be to displace Saddam and his two sons. If you wish to remain unharmed, place your tanks and other armored vehicles in formation and abandon them. Walk away. You and your troops should go home. You and other Iraqi forces will be reconstituted after the regime is changed in Baghdad.

Not surprisingly, many Iraqi officers obeyed the instructions CENTCOM had e-mailed them, on the secret Iraqi network. U.S. troops found many units had neatly parked their tanks in rows outside their bases, thus allowing U.S. aircraft to neatly blow them up. Some Iraqi army commanders sent their troops on leave in the hours before the war. Troops put on civilian clothes and went home, or at least tried to.

Although willing to hack into Iraq's network to engage in a psychological campaign prior to the onset of the conventional attack, the Bush Administration was apparently unwilling to destroy Saddam Hussein's financial assets by cracking into the networks of banks in Iraq and other countries. The capability to do so existed, but government lawyers feared that raiding bank accounts would be seen by other nations as a violation of international law, and viewed as a precedent. The counsels also feared unintended consequences if the U.S. cyber bank robberies hit the wrong accounts or took out entire financial institutions.

The second U.S.-Iraq war, and the more recent Israeli attack on Syria, had demonstrated two uses of cyber war. One use of cyber war is to make a conventional (the U.S. military prefers the term "kinetic") attack easier by disabling the enemy's defenses. Another use of cyber war is to send propaganda out to demoralize the enemy, distributing e-mails and other Internet media in place of the former practice of dropping pamphlets. (Recall the thousands of pieces of paper with instructions in Arabic and stick-figure drawings dropped on Iraqi forces in 1991, telling them how to surrender to U.S. forces. Thousands of Iraqis brought the pamphlets with them when they did surrender.)

The raid on the Syrian nuclear facility and the U.S. cyber activity that preceded the invasion of Iraq are examples of the military using hacking as a tool to assist in a more familiar kind of war. The use of cyberspace by nation-states for political, diplomatic, and military goals does not, however, have to be accompanied by bombing raids or tank battles. A small taste of what a stand-alone cyber war could look like came, somewhat surprisingly, in a little Hanseatic League city of 400,000 people on the shores of the Baltic. The city of Tallinn had become, once again, the capital of an independent Estonia in 1989 when the Soviet Union disintegrated and many of its component republics disassociated themselves from Moscow and the U.S.S.R. Estonia had been forced to become part of the Soviet Union when the Red Army "liberated" the Baltic republic from the Nazis during what the Russians call "the Great Patriotic War."

The Red Army, or at least the Communist Party of the Soviet

Union, didn't want Estonians, or any other East Europeans, to forget the sacrifices that were made "liberating" them. Thus, in Tallinn, as in most East European capitals, they erected one of those giant, heroic statues of a Red Army soldier that the Soviet leaders had such a fondness for. Often these bronzes stood atop the graves of Red Army soldiers. I first stumbled upon such a statue, almost literally, in Vienna in 1974. When I asked the police protecting it why neutral Austria had a giant Communist soldier in its downtown, they told me that the Soviet Union had put it up right after the war and had required the Austrians to promise never to take it down. Indeed, the statue is specifically protected in the treaty the U.S. and Austria signed, along with the Soviets, when American and Soviet troops left Austria in 1950. Back in the 1970s, the Viennese almost uniformly described the enormous bronze as "the only Russian soldier in Vienna who did not rape our women." It seems these statues mean a great deal to the Russians, just as the overseas graves of American World War II dead are sacred ground to many American veterans, their families, and their descendants. The giant bronze statues also had significant meaning to those who were "liberated," but that meaning was something entirely different. The statues and the dead bodies of Red Army soldiers under them were, symbolically, lightning rods. In Tallinn, the statue also attracted cyber lightning.

Tensions between ethnic Russians living in Estonia and the native Estonians themselves had been building ever since the little nation had declared its independence again at the end of the Cold War. The majority of Estonians sought to remove any sign of the five oppressive decades during which they had been forced to be part of the Soviet Union. In February 2007, the legislature passed a Forbidden Structures Law that would have caused anything denoting the occupation to be taken down, including the giant bronze soldier.

Estonians still resented the desecration of their own veterans' graves that had followed the appearance of the Red Army.

Moscow complained that moving the bronze soldier would defame the heroic Soviet dead, including those buried around the giant bronze. Seeking to avoid an incident, the Estonian President vetoed the law. But public pressure to remove the statue grew, just as a Russian ethnic group dedicated to protecting the monument and an Estonian nationalist group threatening to destroy it became increasingly militant. As the Baltic winter warmed into spring, the politics moved to the street. On April 27, 2007, now known as Bronze Night, a riot broke out between radicals from both ethnic factions, with the police and the statue caught in the middle. Authorities quickly intervened and moved the statue to a new, protected location in the military cemetery. Far from quelling the dispute, the move ignited indignant nationalist responses in the Moscow media and in Russia's legislature, the Duma.

This is when the conflict moved into cyberspace. Estonia, oddly, is one of the most wired nations in the world, ranking, along with South Korea, well ahead of the United States in the extent of its broadband penetration and its utilization of Internet applications in everyday life. Those advances made it a perfect target for cyber attack. After Bronze Night, suddenly the servers supporting the most often utilized webpages in Estonia were flooded with cyber access requests, so flooded that some of the servers collapsed under the load and shut down. Other servers were so jammed with incoming pings that they were essentially inaccessible. Estonians could not use their online banking, their newspapers' websites, or their government's electronic services.

What had hit Estonia was a DDOS, a distributed denial of service attack. Normally a DDOS is considered a minor nuisance, not a major weapon in the cyber arsenal. Basically it is a preprogrammed flood of Internet traffic designed to crash or jam networks. It is "distributed" in the sense that thousands, even hundreds of thousands, of computers are engaged in sending the electronic pings to a handful of targeted locations on the Internet. The attacking computers are called a "botnet," a robotic network, of "zombies," computers that are under remote control. The attacking zombies were following instructions that had been loaded onto them without their owners' knowledge. Indeed, the owners usually cannot even tell when their computers have become zombies or are engaged in a DDOS. A user may notice that the laptop is running a little slowly or that accessing webpages is taking a little longer than normal, but that is the only indicator. The malicious activity is all taking place in the background, not appearing on the user's screen. Your computer, right now, might be part of a botnet.

What has happened, often weeks or months before a botnet went on the offensive, is that a computer's user went to an innocent-looking webpage and that page secretly downloaded the software that turned their computer into a zombie. Or they opened an e-mail, perhaps even one from someone they knew, that downloaded the zombie software. Updated antivirus or firewall software may catch and block the infections, but hackers are constantly discovering new ways around these defenses.

Sometimes the zombie computer sits patiently awaiting orders. Other times it begins to look for other computers to attack. When one computer spreads its infection to others, and they in turn do the same, we have the phenomenon known as a "worm," the infection worming its way from one computer through thousands to millions. An infection can spread across the globe in mere hours.

In Estonia the DDOS was the largest ever seen. It appeared that several different botnets, each with tens of thousands of infected machines that had been sleeping, were now at work. At first, the Estonians thought that the takedown of some of their webpages was just an annoyance sent at them from outraged Russians. Then the botnets started targeting Internet addresses most people would not know, not those of public webpages, but the addresses of servers running parts of the telephone network, the credit-card verification system, and the Internet directory. Now over a million computers were engaged in sending a flood of pings toward the servers they were targeting in Estonia. Hansapank, the nation's largest bank, was staggered. Commerce and communications nationwide were being affected. And the attacks did not stop.

In most previous eruptions of a DDOS attack, one site would be hit for a few days. This was something different. Hundreds of key sites in one country were being hit week after week, unable to get back up. As Internet security experts rushed to Tallinn from Europe and North America, Estonia brought the matter before the North Atlantic Council, the highest body of the NATO military alliance. An ad hoc incident response team began trying countermeasures that had been successful in the past with smaller DDOS attacks. The zombies adapted, probably being reprogrammed by the master computers. The attacks continued. Using trace-back techniques, cyber security experts followed the attacking pings to specific zombie computers and then watched to see when the infected machines "phoned home" to their masters. Those messages were traced to controlling machines, and sometimes further traced to higher-level controlling devices. Estonia claimed that the ultimate controlling machines were in Russia, and that the computer code involved had been written on Cyrillic-alphabet keyboards.

The Russian government indignantly denied that it was engaged in cyber war against Estonia. It also refused Estonia's formal diplomatic request for assistance in tracing the attackers, although a standing bilateral agreement required Moscow to cooperate. Informed that the attacks had been traced back to Russia, some government officials admitted that it was possible perhaps that patriotic

Russians, incensed at what Estonia had done, were taking matters into their own hands. Perhaps.

But even if the "patriotic Russians" theory were to be believed, it left unanswered the question of why the Russian government would not move to stop such vigilantism. No one doubted for a minute that the KGB's successors had the ability to find the culprits and to block the traffic. Others, more familiar with modern Russia, suggested that what was at work was far more than a passive Russian police turning a blind eye to the hooliganism of overly nationalistic youth. The most adept hackers in Russia, apart from those who are actual government employees, are usually in the service of organized crime. Organized crime is allowed to flourish because of its unacknowledged connection to the security services. Indeed, the distinction between organized criminal networks and the security services that control most Russian ministries and local governments is often blurry. Many close observers of Russia think that some senior government officials permit organized crime activity for a slice of the profits, or, as in the case of Estonia, for help with messy tasks. Think of Marlon Brando as the Godfather saying, "Someday . . . I will call upon you to do a service for me . . ."

After Bronze Night, the Russian security services had encouraged domestic media outlets to whip up patriotic sentiment against Estonia. It is not a stretch to imagine that they also asked organized crime groups to launch the hackers in their employ, perhaps even giving those hackers some information that would prove helpful. Did the Russian government security ministries engage in cyber attacks on Estonia? Perhaps that is not the right question. Did they suggest the attacks, facilitate them, refuse to investigate or punish them? And, in the end, does the distinction really matter when you are an Estonian unable to get your money out of a Hansapank ATM?

. . .

Following the cyber attack, NATO moved to create a cyber defense center. It opened in 2008, a few miles from the site where the giant bronze solider had originally stood. On the original site of the bronze soldier there is a nice little grove of trees now. Unfortunately, the NATO center in Tallinn was of little use when another former Soviet satellite republic, Georgia, and Mother Russia got into a tussle over some small disputed provinces.

The Republic of Georgia lies directly south of Russia along the Black Sea, and the two nations have had a decidedly unequal relationship for well over a century. Georgia is geographically slightly smaller than the state of South Carolina and has a population of about four million people. Given its location and size, Georgia has been viewed by Moscow as properly within the Kremlin's "sphere of influence." When the original Russian empire began to disintegrate after the Russian Revolution, the Georgians tried to make a break for it while the Russians were too busy fighting each other, declaring Georgian independence in 1918. As soon as the Russians finished fighting each other, however, the victorious Red Army quickly invaded Georgia, installed a puppet regime, and made Georgia part of the Union of Soviet Socialist Republics. Soviet control of Georgia lasted until 1991, when, as the central Russian government was again in turmoil, Georgia once more took the opportunity to declare independence.

Two years later, Georgia lost control of two territories, South Ossetia and Abkhazia. Supported by Moscow, the local Russian populations in those territories succeeded in defeating the ragtag Georgian army and expelling most Georgians. The territories then set up "independent" governments. Although still legally part of Georgia as far as the rest of the world was concerned, the regions relied on Russian funding and protection. Then, in July 2008, South

Ossetian rebels (or Russian agents, depending upon whose version of events you trust) provoked a conflict with Georgia by staging a series of missile raids on Georgian villages.

The Georgian army, predictably, responded to the missile strikes on its territory by bombing the South Ossetian capital city. Then, on August 7, Georgia invaded the region. Not surprised by this turn of events, the Russian army moved the next day, quickly ejecting the Georgian army from South Ossetia. Precisely at the same time that the Russian army moved, so did its cyber warriors. Their goal was to prevent Georgians from learning what was going on, so they streamed DDOS attacks on Georgian media outlets and government websites. Georgia's access to CNN and BBC websites were also blocked.

In the physical world, the Russians also bombed Georgia and took over a small chunk of Georgian territory that was not in dispute, allegedly to create a "buffer zone." While the Georgian army was busy getting routed in Ossetia, rebel groups in Abkhazia decided to take advantage of the situation and push out any remaining Georgians, with a little help from their Russian backers. The Russian army then took another little slice of Georgian land, as an additional buffer. Five days later, most of the fighting was over. French President Nicolas Sarkozy brokered a peace agreement in which the Russians agreed to withdraw from Georgia immediately and to leave the disputed territories once an international peacekeeping force arrived to fill the security vacuum. That force never arrived, and within a few weeks Russia recognized South Ossetia and Abkhazia as independent states. The declared independent states then invited their Russian benefactors to stay.

To most in the U.S., except then presidential candidate John McCain, who tried to portray it as a national security crisis for America, all of this activity in Georgia seemed remote and unimportant. As soon as most Americans reassured themselves that the news reports they heard about the invasion of Georgia did not really mean Russian army troops or General Sherman again marching on Atlanta, they tuned out. The event's true significance, beyond what it revealed of the Russian rulers' thinking about their former empire, lies in what it exposed of their attitudes toward the use of cyber attacks.

Before fighting broke out in the physical world, cyber attacks hit Georgian government sites. In the initial stages, the attackers conducted basic DDOS attacks on Georgian government websites and hacked into the web server of the President's site to deface it, adding pictures that compared the Georgian leader, Mikheil Saakashvili, to Adolf Hitler. It had seemed trivial, even juvenile, at first. Then the cyber attacks picked up in intensity and sophistication just as the ground fighting broke out.

Georgia connects to the Internet through Russia and Turkey. Most of the routers in Russia and Turkey that send traffic on to Georgia were so flooded with incoming attacks that no outbound traffic could get through. Hackers seized direct control of the rest of the routers supporting traffic to Georgia. The effect was that Georgians could not connect to any outside news or information sources and could not send e-mail out of the country. Georgia effectively lost control of the nation's ".ge" domain and was forced to shift many government websites to servers outside the country.

The Georgians tried to defend their cyberspace and engage in "work-arounds" to foil the DDOS attack. The Russians countered every move. Georgia tried to block all traffic coming from Russia. The Russians rerouted their attacks, appearing as packets from China. In addition to a Moscow-based master controller for all the botnets being used in the attacks, servers in Canada, Turkey, and, ironically, Estonia were also used to run botnets.

Georgia transfered the President's webpage to a server on Google's blogspot in California. The Russians then set up mock presidential sites and directed traffic to them. The Georgian banking sector shut down its servers and planned to ride out the attacks, thinking that a temporary loss of online banking was a better bargain than risking the theft of critical data or damage to internal systems. Unable to get to the Georgian banks, the Russians had their botnets send a barrage of traffic to the international banking community, pretending to be cyber attacks *from* Georgia. The attacks triggered an automated response at most of the foreign banks, which shut down connections to the Georgian banking sector. Without access to European settlement systems, Georgia's banking operations were paralyzed. Credit card systems went down as well, followed soon after by the mobile phone system.

At their peak, the DDOS attacks were coming from six different botnets using both computers commandeered from unsuspecting Internet users and from volunteers who downloaded hacker software from several anti-Georgia websites. After installing the software, a volunteer could join the cyber war by clicking on a button labeled "Start Flood."

As in the Estonian incident, the Russian government claimed that the cyber attacks were a populist response that was beyond the control of the Kremlin. A group of Western computer scientists, however, concluded that the websites used to launch the attacks were linked to the Russian intelligence apparatus. The level of coordination shown in the attacks and the financing necessary to orchestrate them suggest this was no casual cyber crusade triggered by patriotic fervor. Even if the Russian government were to be believed (namely, that the cyber storm let loose on Georgia, like the previous one on Estonia, was not the work of its official agents), it is very clear that the government did nothing to stop it. After all, the huge Soviet intelligence agency, the KGB, is still around, although with a slightly different organizational structure and name. Indeed the KGB's power has only increased under the regime of its alumnus, Vladimir Putin. Any large-scale cyber activity in Russia, whether done by government, organized crime, or citizens, is done with the approval of the intelligence apparatus and its bosses in the Kremlin.

If it was, as we suspect, effectively the Russian government that asked for the "vigilante" DDOS and other cyber attacks as a stand-alone punishment of Estonia and later conducted them as an accompaniment to kinetic war on Georgia, those operations do not begin to reveal what the Russian military and intelligence agencies could do if they were truly on the attack in cyberspace. The Russians, in fact, showed considerable restraint in the use of their cyber weapons in the Estonian and Georgian episodes. The Russians are probably saving their best cyber weapons for when they really need them, in a conflict in which NATO and the United States are involved.

For years U.S. intelligence officials had thought that if any nation were going to use cyber weapons, even in the small ways demonstrated in Estonia and Georgia, the likely first movers would be Russia, China, Israel, and, of course, the United States. The nation that joined that club in the summer of 2009 came as a surprise to some.

It was a little after seven p.m. in Reston, Virginia, on the last Monday in May 2009. Outside, the rush-hour traffic was beginning to thin on the nearby Dulles Airport Access Road. Inside, a flat screen at the U.S. Geological Survey had just indicated a 4.7 magnitude earthquake in Asia. The seismic experts began narrowing in on the epicenter. It was in the northeastern corner of the Korean Peninsula, specifically forty-three miles from a town on the map called Kimchaek. The data showed that there had been a similar event very nearby in October 2006. That one had turned out to be a nuclear explosion. So did this one.

After years of negotiating with the U.S., as well as with China and Russia, the weird, hermetic government of North Korea had decided to defy international pressure and explode a nuclear bomb, for the second time. Their first attempt, three years earlier, had been characterized by some Western observers as something like a "partial fizzle." In the ensuing hours after this second blast, U.S. Ambassador to the United Nations Susan E. Rice was attached to the phone in her suite at New York's Waldorf Towers. She consulted with the White House and the State Department, then she began to call other UN ambassadors, notably the Japanese and South Koreans. The South Korean who is the head of the UN, Secretary General Ban Ki-moon, agreed to an emergency meeting of the Security Council. The outcome of that feverish round of diplomatic consultations was, eventually, further international condemnation of North Korea and further sanctions on the impoverished tyranny. A decade and a half's worth of diplomacy to prevent a North Korean nuclear capability had come to naught. Why?

Some observers of the Pyongyang government explained that the destitute North had no other leverage to extract concessionary loans, free food, and gifts of oil. It had to keep selling the same thing over and over, a promise not to go further with its nuclear capability. Others pointed to the rumored ill health of the strange man known in the North as the Dear One, Kim Jong-il, the leader of the Democratic People's Republic of Korea. The tea-leaf readers believed that the Dear One knew that he was fading and had selected Number Three Son, Kim Jong-un, a twenty-five-year-old, to succeed him. To prevent the United States, or South Korea, from taking advantage of the transition period, the analysts claimed, the North believed it had to rattle its sabers, or at least its atoms. The pattern with North Korea in the past had been to threaten, get attention, give a taste of what awful things might happen, then offer to talk, and eventually to cut a deal to enrich their coffers.

If the detonation was designed to provoke the United States and others to rush with offers of wheat and oil, it failed. Having condemned the explosion and announced the movement of defensive missiles to Hawaii, as June moved on, the U.S. leadership shifted its focus back to health care reform, Afghanistan, and self-flagellation over its own intelligence activities. Somewhere in the bureaucracy an American official publicly announced that the U.S. would again be conducting a cyber war exercise known as Cyber Storm to test the defense of computer networks. The 2009 exercise would involve other nations, including Japan and Korea, the one in the south. North Korean media soon responded by characterizing the pending exercise as a cover for an invasion of North Korea. That kind of bizarre and paranoid analysis is par for the course with North Korea. No one in Washington thought twice about it.

As the July 4 break began in Washington, bureaucrats scattered to vacation homes on East Coast beaches. Tourists in Washington swarmed to the National Mall, where a crowd of several hundred thousand watched the "rockets' red glare" of a sensational fireworks display, a signature of the Fourth of July holiday. On the other side of the world, the association of rockets and the Fourth was not lost on some in the North Korean leadership. In outer space, a U.S. satellite detected a rocket launch from North Korea. Computers in Colorado quickly determined that the rocket was short-ranged and was fired into the sea. Then there was another rocket launch. Then another and another. Seven North Korean rockets were fired on the Fourth of July. Whether a plea for help, or more saber rattling, it certainly seemed like a cry for attention. But that cry did not stop there. It moved into cyberspace.

Right before the Fourth of July holiday, a coded message was sent out by a North Korean agent to about 40,000 computers around the world that were infected with a botnet virus. The message contained a simple set of instructions telling the computer to start pinging a list of U.S. and South Korean government websites and international companies. Whenever the infected computers were turned on, they silently joined the assault. If your computer was one of the zombies, you might have noticed your processor was running slowly and your Web requests were taking a bit longer to process, but nothing too out of the ordinary. Yes, it was another DDOS attack by zombies in a botnet. At some time over the weekend, the U.S. government did notice when dhs.gov and state.gov became temporarily unavailable. If anyone actually thought of consulting the Department of Homeland Security terrorist threat level before deciding to go watch the fireworks on the National Mall, they would not have been able to gain that information from the Department of Homeland Security's website.

Each of those zombie computers was flooding these sites with requests to see their pages in another distributed denial of service attack. The U.S. websites were hit with as many as 1 million requests per second, choking the servers. The Treasury, Secret Service, Federal Trade Commission, and Department of Transportation web servers were all brought down at some point between July 4 and July 9. The NASDAQ, New York Mercantile, and New York Stock Exchange sites were also hit, as was the *Washington Post*. The DDOS aimed at the White House failed, however. To prevent the first DDOS attack against the White House in 1999, I had arranged with a company known as Akamai to route traffic seeking the White House website to the nearest of over 20,000 servers scattered around the world. When the Korean attack hit in 2009, the DDOS went to the White House servers nearest the source of the attacker. Thus, only sites hosting the White House website in Asia had trouble. White House spokesperson Nick Shapiro apologized in a halfhearted way to any web surfers in Asia who might not have been able to get onto the White House site. Then the second and third waves hit.

Another 30,000 to 60,000 computers infected with a different variant of the virus were told to target a dozen or more South Korean government sites, Korean banks, and a South Korean Internet security company on July 9. The attackers were apparently convinced that the attacks on U.S. sites were no longer going to be effective after the government and major corporations began working with Internet service providers (ISPs) to filter out the attacks. At 6:00 p.m. Korea time on July 10, the final assault began. The now estimated 166,000 computers in seventy-four countries started flooding the sites of Korean banks and government agencies.

Ultimately, the damage was contained. The attack did not attempt to gain control of any government systems, nor did it disrupt any essential services. But it was likely only meant as a shot across the bow. What we do know is that there was an agenda and motivation for the attack. This was not a worm simply released into the wilds of the Internet and allowed to propagate. Someone controlled and directed the attack and modified its target list to focus on the more vulnerable Korean sites.

The U.S. government has yet to directly attribute the attack to North Korea, though South Korea has not been shy about doing so. The timing of the attacks does suggest the North Korean regime is the prime suspect, but definite attribution is difficult. The infected computers attempted to contact one of eight "command and control servers" every three minutes. These servers sent instructions back to the infected zombie computers, telling them which websites to attack. The eight masters were in South Korea, the United States, Germany, Austria, and, interestingly, Georgia (the country).

The Korea Communications Commission has endorsed the judgment of a Vietnamese firm, Bach Khoa Internetwork Security (BKIS), that these eight servers were controlled from a server in Brighton, England. From there, the trail goes cold, though it does not look like the mastermind behind the attack was sitting in front of a keyboard near the beach in Brighton. South Korea's National Intelligence Service (NIS) suspects that a North Korean military research institute set up to destroy South Korea's communications infrastructure was involved. The NIS said in a statement following the attack that it had evidence that pointed to North Korea.

The NIS maintains that the North Korean hacker unit, known as Lab 110, or the "technology reconnaissance team," was ordered to prepare a plan for cyber attack on June 7. That order directed the unit to "destroy the South Korean puppet communications networks in an instant," following the decision by the South Koreans to participate in Excercise Cyber Storm. The North called the exercise "an intolerable provocation as it revealed ambition to invade the DPRK."

South Korea is now preparing for all-out cyber war with the North. Just before the attacks began, South Korea had announced plans for establishing a cyber warfare command by 2012. After the attacks, it sped up the timeline to January 2010. What the South's new cyber warfare command will do the next time the North attacks in cyberspace is unclear.

If North Korea attacks in cyberspace again, options for responding are relatively limited. Sanctions cannot be made much tighter. Suspended food aid cannot be suspended further. Any military action in retaliation is out of the question. The 23 million residents of metropolitan Seoul live within range of North Korea's artillery pieces, set along the demilitarized zone in what military planners refer to as "the kill box."

There is also little possibility of responding in kind, since North Korea has little for either U.S. or South Korean cyber warriors to attack. In 2002, Donald Rumsfeld and other Bush Administration officials advocated the invasion of Iraq because Afghanistan was not a "target rich" environment, with not enough military hardware, bases, or major infrastructure for the U.S. to blow up. North Korea is the cyber equivalent of Afghanistan.

Nightearth.com compiled satellite photos of the planet at night taken from space. Its composite map shows a well-lit planet. South

Korea looks like a bright island separated from China and Japan by the sea. What looks like the sea, the Korean peninsula north of Seoul, is almost completely dark. North Korea barely has an electric grid. Fewer than 20,000 of North Korea's 23 million citizens have cell phones. Radios and TVs are hardwired to tune only into official government channels. And as far as the Internet is concerned, the *New York Times*'s judgment from 2006 that North Korea is a "black hole" still stands. *The Economist* described the country as "almost as cut off from the virtual world as it is from the real one." North Korea operates about thirty websites for external communication with the rest of the world, mostly to spread propaganda about its neighbor to the south. A handful of Western hotels are permitted satellite access, and North Korea does run a limited internal network for a few lucky citizens who can go to the Dear One's website, but almost nowhere else.

While North Korea may not have invested much in developing an Internet infrastructure, it has invested in taking down the infrastucture in other countries. Unit 110, the unit suspected of carrying out the July cyber attacks, is only one of North Korea's four cycle warfare units. The Korean People's Army (KPA) Joint Chiefs Cyber Warfare Unit 121 has over 600 hackers. The Enemy Secret Department Cyber Psychological Warfare Unit 204 has 100 hackers and specializes in cyber elements of information warfare. The Central Party's Investigations Department Unit 35 is a smaller but highly capable cyber unit with both internal security functions and external offensive cyber capabilities. Unit 121 is by far the largest and, according to one former hacker who defected in 2004, the best trained. The unit specializes in disabling South Korea's military command, control, and communications networks. It has elements stationed in China because the Internet connections in North Korea are so few and so easily identified. Whether the Beijing government knows the full extent of the North Korean presence and activity is unclear, but few things escape China's secret police, particularly on the the Internet. One North Korean cyber war unit is reportedly located at the Shanghai Hotel in the Chinese town of Dandong, on the North Korean border. Four floors are allegedly rented out to Unit 110 agents. Another unit is in the town of Sunyang, where North Korean agents have reportedly rented out several floors in the Myohyang Hotel. Agents have apparently been spotted moving fiber-optic cables and state-of-the-art computer network equipment into these properties. All told, North Korea may have from 600 to 1,000 KPA cyber warfare agents acting in cells in the PRC, under a commander with the rank of Lieutenant Colonel. North Korea selects elite students at the elementary-school level to be groomed as future hackers. These students are trained on programming and computer hardware in middle and high school, after which they automatically enroll at the Command Automation University in Pyongyang, where their sole academic focus is to learn how to hack into enemy network systems. Currently 700 students are reportedly enrolled. They conduct regular cyber warfare simulated exercises against each other, and some infiltrate Japan to learn the latest computer skills.

The July 2009 attack, though not devastating, was fairly sophisticated. The fact that it was controlled and not simply released to do damage indiscriminately shows that the attackers knew what they were doing. The fact that it lasted for so many days is also a testament to the effort put into propagating the virus from several sources. These attributes suggest that the attack was not the work of some teenagers with too much time on their hands. Of course, North Korea sought "deniability," creating sufficient doubt about who did the attack so that they could claim it was not them.

While researchers have found that part of the program was written using a Korean-language web browser, that would just as likely implicate South Korean hackers for hire, of which there are many in that highly wired nation. These same researchers, however, are troubled by the fact that the code writer didn't try to disguise its Korean origin. Someone sophisticated enough to write the code should also have been sophisticated enough to cover his or her tracks. Perhaps whoever ordered the code written wanted that clue to be found.

The South Korean government and many analysts in the United States concluded that the person who ordered the attack was the Dear One, and that he had demonstrated North Korea's strength in cyberspace at the same time that he had done so with the rocket barrage. The message was: I am still in charge and I can make trouble with weapons that can eliminate your conventional superiority. Having sent that message, a few weeks later North Korean diplomats offered an alternative. They were prepared to talk, even to free two American prisoners. Shortly thereafter, in a scene reminiscent of the movie *Team America: World Police*, Bill Clinton was sitting down with the Dear One. Unlike the marionette portraying UN nuclear inspector Hans Blix in the movie, Clinton did not drop through a trapdoor into a shark tank, but it seemed likely that North Korea had placed trapdoors on computer networks on at least two continents.

Months after the July 2009 North Korean cyber activity, Pentagon analysts concluded that the purpose of the DDOS attacks may have been to determine what level of botnet activity from South Korea would be sufficient to jam the fiber-optic cables and routers leading out of the country. If North Korean agents in South Korea could flood the connection, they could effectively cut the country off from any Internet connection to the rest of the world. That would be valuable for the North to do in a crisis, because the U.S. employs those connections to coordinate the logistics of any U.S. military reinforcements. The North Korean preparation of the cyber battlefield continued. In October, three months after the DDOS attacks, South Korean media outlets reported that hackers had infiltrated the Chemicals Accident Response Information System and had withdrawn a significant amount of classified information on 1,350 hazardous chemicals. The hackers, believed to be North Koreans, obtained access to the system through malicious code implanted in the computer of a South Korean army officer. It took seven months for the South Koreans to discover the hack. North Korea now knows how and where South Korea stores its hazardous gases, including chlorine used for water purification. When chlorine is released into the atmosphere, it can cause death by asphyxiation, as demonstrated horribly on the battlefields of World War I.

The new "cyber warriors" and much of the media herald these incidents as the first public clashes of nation-states in cyberspace. There are other examples, including operations by China, Taiwan, Israel, and others. Some have called the Estonian case "WWI", that is, Web War One.

Others look at these and other recent incidents and do not see a new kind of warfare. They see in the Israeli attack a new form of airborne electronic jamming, something that has been happening in other ways for almost half a century. The American actions in Iraq appear to these doubters to be marginal and mainly propaganda. In the Russian and North Korean activities the doubters see only harassment and nuisance-value disruption.

Of course, the Syrians, Iraqis, Estonians, Georgians, and South Koreans saw these events as far more than a nuisance. I tend to agree. I have walked through these recent, well-known cyber clashes mainly to demonstrate that nation-state conflict involving cyber attacks has begun. Beyond that incontestable observation, however, there are five "take-aways" from these incidents:

Cyber war is real. What we have seen so far is far from indicative of what can be done. Most of these well-known skirmishes in cyberspace used only primitive cyber weapons (with the notable exception of the Israeli operation). It is a reasonable guess that the attackers did not want to reveal their more sophisticated capabilities, yet. What the United States and other nations are capable of doing in a cyber war could devastate a modern nation.

Cyber war happens at the speed of light. As the photons of the attack packets stream down fiber-optic cable, the time between the launch of an attack and its effect is barely measurable, thus creating risks for crisis decision makers.

Cyber war is global. In any conflict, cyber attacks rapidly go global, as covertly acquired or hacked computers and servers throughout the world are kicked into service. Many nations are quickly drawn in.

Cyber war skips the battlefield. Systems that people rely upon, from banks to air defense radars, are accessible from cyberspace and can be quickly taken over or knocked out without first defeating a country's traditional defenses.

Cyber war has begun. In anticipation of hostilities, nations are already "preparing the battlefield." They are hacking into each other's networks and infrastructures, laying in trapdoors and logic bombs—now, in peacetime. This ongoing nature of cyber war, the blurring of peace and war, adds a dangerous new dimension of instability.

As later chapters will discuss, there is every reason to believe that most future kinetic wars will be accompanied by cyber war, and that other cyber wars will be conducted as "stand-alone" activities, without explosions, infantry, airpower, and navies. There has not yet, however, been a full-scale cyber war in which the leading nations in this kind of combat employ their most sophisticated tools against each other. Thus, we really do not know who would win, nor what the results of such a cyber war would be. This book will lay out why the unpredictability associated with full-scale cyber war means that there is a credible possibility that such conflict may have the potential to change the world military balance and thereby fundamentally alter political and economic relations. And it will suggest ways to reduce that unpredictability.

CHAPTER TWO

CYBER WARRIORS

In a television ad, a crew-cut young man in a jumpsuit walks around a darkened command center, chatting with subordinates who are illuminated by the greenish light from their computer screens. We hear his voice over the video: "control of power systems . . . water systems . . . that is the new battlefield . . . in the future this is going to be the premier war-fighting domain . . . this is going to be where the major battles are fought." He then looks right at the camera and says, "I am Captain Scott Hinck, and I am an Air Force Cyber Warrior." The screen fades to black, and then three words appear: " Air, Space, Cyberspace." Then, as the ad ends, we see a winged symbol and the name of the sponsor, "United States Air Force."

So now we know what one cyber warrior looks like. At least in

Scott's case, he looks a lot like the bright, fit, earnest officers who populate the world's most potent military. That is not quite our image of hackers, whom movies have portrayed as acned, disheveled guys with thick glasses. To attract more of those with the skills needed to understand how to fight cyber war, however, the Air Force seems to think it may have to bend the rules. "If they can't run three miles with a pack on their back, but they can shut down a SCADA system," mused Air Force Major General William Lord, "we need to have a culture where they can fit in." (A SCADA system is the software that controls networks such as electric power grids.) That progressive attitude reflects the U.S. Air Force's strong desire to play the leading role for the U.S. in cyber war. That service was the first to create an organization for the purpose of combat in the new domain: U.S. Air Force Cyber Command.

THE FIGHT FOR CYBER WAR

In October 2009, when the doors opened on the multiservice, joint U.S. Cyber Command, the Navy had already followed the Air Force in standing up its own cyberwarfare unit. All the new organizations and big pronouncements gave some the impression that the U.S. military was just getting interested in cyber warfare, coming rather late to the game. Not so. The Department of Defense invented the Internet, and the possibility of using it in warfare was not overlooked even in its early days. As highlighted above, in chapter 1, early cyber warriors had a plan back in the first Gulf War to use cyber weapons to take down Iraq's air defense system. Shortly after that war, the Air Force set up its Info War Center. In 1995, National Defense University graduated its first class of officers trained to lead cyber war campaigns.

Some in the 1990s military did not fully understand what cyber war meant and thought of it as "info ops," part of psychological warfare, or "psyops" (using propaganda to influence the outcome of wars). Others, particularly those in the intelligence branches, were seeing the ever expanding Internet as a bonanza for electronic espionage. It started to become pretty obvious that once you had penetrated a network to collect information, a few more keystrokes could take that network down.. As this realization grew among the electronic intelligence officers, they had a dilemma. The intelligence guys knew that if they told the "operators" (the fighting units) that the Internet was making a new kind of war possible, they would lose some control of cyberspace to the "warriors." On the other hand, the warriors would still have to rely on the intelligence geeks to do anything in cyberspace. Moreover, the opportunities cyberspace offered to relatively easily do significant damage to an enemy were too good to pass up. Slowly, the warriors realized that the geeks were on to something.

By the time George W. Bush was starting his second term, the importance of cyber war to the Pentagon became apparent, as the Air Force, Navy, and intelligence agencies engaged in a bitter struggle to see who would control this new area of warfare. Some advocated the creation of a Unified Command, bringing the units of all three services under one integrated structure. There were already Unified Commands for transportation, strategic nuclear war, and for each of the world's regions. When it appeared in the early 1980s that there would be a large role for the military in outer space, the Pentagon created a Unified Command for what it then thought of as a new domain for war-fighting, a domain that the United States had to control. U.S. Space Command lasted from 1985 to 2002, by which time it had become clear that neither the U.S. nor any other government had the money to do much in space. Space Command was folded into Strategic Command (STRATCOM), which operates the strategic nuclear forces. STRATCOM, headquartered at a bomber base in Nebraska, was also given the centralized responsibility for cyber war in 2002. The Air Force, however, was set on running the actual war-fighting units. The creation of Air Force Cyber Command and the standing given to cyberspace in the Air Force recruitment ads jarred the other services and many in the Pentagon.

Some were concerned that the Air Force was talking too openly about something they believed should have been kept secret: the mere existence of cyber war capability. Yet there was the civilian Air Force Secretary (a vestigial post from the time before there was a strong civilian Defense Department) saying publicly, "Tell the nation the age of cyber war is here." There were those damn ads, including one that said, ominously, that in the future a blackout "could be a cyber attack." Another ad showed the Pentagon and claimed that it was "attacked" millions of times a day in cyberspace, but it was defended by the likes of an Air Force sergeant shown at his keyboard. There were persistent interviews and speeches by Air Force leaders who sounded very aggressive about their intentions. "Our mission is to control cyberspace, both for attacks and defense," Lieutenant General Robert Elder had admitted. The Director of the Air Force Cyberspace Operations Task Force had been equally candid: "If you are defending in cyberspace, you're already too late. If you do not dominate in cyberspace, you cannot dominate in other domains. If you are a developed country [and you are attacked in cyberspace], your life comes to a screeching halt."

By 2008, those in the Pentagon not wearing blue uniforms had become persuaded about the importance of cyber war, but they were also convinced that it should not just be conducted by the Air Force. An integrated multiservice structure was agreed on in principle, but many were reluctant to "make the Space Command mistake again." They did not want to create a Unified Command for what might prove to be a passing fad, as war fighting in space had been. The compromise was that a multiservice Cyber Command would be created, but it would remain subordinated to STRATCOM, at least on paper. The Air Force would have to stop calling its organization a command and would instead have to be satisfied with a "numbered air force," their basic organizational unit, like Navy's numbered fleets. The agreement in principle did not resolve all of the major issues standing in the way of a new command.

The intelligence community had a view. Under the post-9/11 reorganization, there was now a single person in charge of all eighteen U.S. intelligence agencies. In 2008, that man was Mike McConnell. He looked much the part of what he had recently been, a well-to-do businessman often seen in the halls of Wall Street financial institutions. He had come to the intelligence job from the global consulting giant Booz Allen Hamilton. Slightly hunched over and wearing thick glasses, the soft-spoken McConnell had not taken a traditional path to leadership at Booz. For most of his life, he had been in Navy intelligence, retiring as a three-star (or vice) admiral, the man in charge of the world's premier electronic intelligence organization, the National Security Agency (NSA).

Hearing McConnell, or his successor, Air Force General Ken Minihan, talk about NSA even on an unclassified basis, you begin to understand why they believe re-creating some of its capabilities elsewhere is folly and perhaps impossible. They both speak with real reverence about the decades of experience and expertise NSA has in "doing the impossible" when it comes to electronic espionage. NSA's involvement in the Internet grew out of its mission to listen to radio signals and telephone calls. The Internet was just another electronic medium. As Internet usage grew, so did intelligence agencies' interest in it. Populated with Ph.D.s and electrical engineers, NSA quietly became the world's leading center of cyberspace expertise.

Although not authorized to alter data or engage in disruption and damage, NSA thoroughly infiltrated the Internet infrastructure outside of the U.S. to spy on foreign entities.

When McConnell left NSA in 1996 for Booz Allen Hamilton, he continued his focus on the Internet, working with leading U.S. companies on their cyber security plans for over a decade. Returning to the spook business in 2007, he tried, as the second-ever Director of National Intelligence, to assert authority over all of the U.S. intelligence agencies, including CIA. In doing so, his long-standing friendship with CIA Director Mike Hayden was damaged. Hayden had also once been Director of NSA, or as they say it in the intelligence community DIRNSA (pronounced "*dern*-sah"). Hayden remained an active-duty four-star Air Force General much of the time he ran CIA.

Because both Mikes (McConnell and Hayden) had the background of running NSA, they agreed on at least one thing: any new Cyber Command must not try to replicate the capabilities it had taken decades to develop at NSA. If anything were to be done, they and many of the other NSA alumni believed, NSA should just *become* the new Cyber Command. Their views mattered in the Pentagon, since they were, or had been, senior military officers, and they actually knew something about cyberspace. To counter the "NSA takeover" of Cyber Command, some in the military argued that NSA was really a civilian organization, an intelligence unit, and therefore could not legally fight wars. They talked about "Title 50 versus Title 10" authority, referring to the parts of the U.S. Code that give legal authority and limitations to various government departments and agencies. Such laws can, of course, be changed if they have outlived their utility. Nonetheless, the issue of who would run America's cyber wars soon became a battle between military and civilian government lawyers.

In any other alignment of leaders, the outcome would likely have been decided in the military's favor and some new organization would have been built from the ground up, replicating the hacking skills at which NSA was the past master. In 2006, however, the turf-grabbing Secretary of Defense, Donald Rumsfeld, had been replaced after devastating midterm election losses brought on in part by mismanagement of the Iraq War. Rumsfeld's replacement was the president of Texas A&M University, Robert Gates. At the time of his nomination I had known Bob for the better part of three decades and expected that he would be an unusually good Secretary of Defense. He was not a Pentagon man, had not grown up there. Nor was he a national security novice from industry or academia, the type easily manipulated by experienced Pentagon hands. Bob had been a career CIA officer who worked his way up to CIA Director, stopping off in the White House National Security Council along the way. Gates saw the Cyber Command debate from an intelligence community perspective and, more important, from the unique perch one has at the White House. When you are working directly for whoever the President may be at the time, you suddenly realize that there is a national interest that surpasses the turf concerns of whatever bureaucracy you may have come from. Gates had that broader view, and he was a pragmatist.

What resulted was a compromise in which the Director of NSA would become a four-star general (up from three stars) and would also be the head of U.S. Cyber Command. The Pentagon calls having two jobs being "dual hatted." For now, at least, Cyber Command would be a "sub-Unified Command" under STRATCOM. The assets of NSA would be available to support U.S. Cyber Command, thus obviating the need for reinventing many wheels. The Air Force, Navy, and Army would continue to have cyber war units, but they would be run by U.S. Cyber Command. Technically, it would be those war-fighting military units that would actually engage in cyber combat and not the partially civilian intelligence agency that is NSA. While NSA has a lot of expertise in network penetration, under U.S. law (Title 10) the agency is restricted to collecting information and prohibited from war-fighting. Therefore it will have to be military personnel under Title 50 that enter the keystrokes to take down enemy systems. To assist Cyber Command in its defensive role of protecting Defense Department networks, the Pentagon would also co-locate its own Internet service provider at Fort Meade, Maryland, alongside NSA. The Pentagon's ISP is unlike any other, since it runs two of the largest networks in the world. Called the Defense Information Systems Agency (DISA), it is run by a three-star general. Thus, ninety-two years after it opened as an Army base, home to hundreds of horses, Fort Meade became the heart of America's defensive and offensive cyber war forces. Defense contractors are building offices nearby in the hopes of sharing in some of the billions of dollars that will be flowing to Fort Meade. Maryland-area universities are already recipients of large research grants from the nearby military campus, referred to throughout Washington simply as "The Fort."

As a result of the decision to create U.S. Cyber Command, what had been Air Force Cyber Command became the 24th Air Force, with headquarters at Lackland Air Force Base in Texas. This numbered air force won't have any aircraft. The mission of the 24th will be to provide "combat-ready forces trained and equipped to conduct sustained cyber operations, fully integrated within air and space operations." The 24th Air Force will have control of two existing "wings," the 688th Information Operations Wing, formerly the Air Force Information Operations Center, and the 67th Network Warfare Wing, as well as control of a new wing, the 689th Combat Communications Wing. The 688th IOW, as the Information Operations Wing is known, will act as the Air Force's "center of excellence" in cyber operations. The 688th will be a forward-looking element with the mission of finding new ways to create an advantage for the U.S. Air Force using cyber weapons. The 67th Wing will have the day-to-day responsibility for defending Air Force networks and for attacking enemy networks. All totaled, the 24th Air Force will comprise some 6,000 to 8,000 military and civilian cyber warriors.

In case the U.S. Air Force is ever given the order to do as one of its ads suggests ("A power blackout is just a blackout. But in the future, it could be a cyber attack."), the mission will likely fall to the Fighting 67th. Their motto, from pre-cyber days as an aerial reconnaissance outfit, is Lux Ex Tenebris (Light from Darkness). Perhaps they will soon modify it to Tenebra Ex Luce. Despite the demotion of their command, the Air Force lost little of their zeal for cyber war. In the summer of 2009, the head of the U.S. Air Force, General Norton Schwartz, wrote to his officers that "cyberspace is vital to today's fight and to the future U.S. military advantage [and] it is the intent of the United States Air Force to provide a full spectrum of cyberspace capabilities. Cyberspace is a contested domain, and the fight is on—today."

Not to be outdone, the U.S. Navy also reorganized. The Chief of Naval Operations, Admiral Gary Roughead (really), gave himself a new Deputy for Information Dominance. It's not just Roughead and his sailors who are into dominance; the U.S. military in general repeatedly characterizes cyberspace as something to be dominated. It is reminiscent of the Pentagon's way of speaking of nuclear war in the 1960s. The historian of nuclear strategy Lawrence Freedman noted that William Kaufmann, Henry Kissinger, and other strategists realized that there was a need then "to calm the spirit of offense, potent in Air Force circles . . . [whose] rhetoric encouraged a view of war that was out-moded and dangerous." That same sort of macho rhetoric is strong in Air Force cyber war circles today, and apparently in the Navy as well.

Admiral Roughead created not just a Dominance office on the

Navy Staff, but a new "war-fighting" command. The 5th Fleet sails the Arab Gulf, the 6th Fleet the Mediterranean, and the 7th the China Sea. To fight cyber war, the U.S. Navy has reactivated its 10th Fleet. Originally, a small organization during World War II that coordinated antisubmarine warfare in the Atlantic, the 10th Fleet was disbanded shortly after victory over Germany in 1945. Then as now, the 10th Fleet was a "paper" or "phantom" fleet that had no ships. It was a land-based organization that filled a necessary coordinating role. Modest in scope and scale, the 10th Fleet in World War II served its limited purpose well with no more than fifty intelligence officers. This time, the Navy has much more ambitious plans for the 10th Fleet. The existing Naval Network Warfare Command, known as NETWARCOM, will continue its operational responsibilities subordinated to the 10th Fleet. Although the Navy has not done the sort of public self-promotion of its cyber warriors that the Air Force has, they insist that they have as much tech savvy as "the fly boys." Perhaps to prove that point, one Naval officer told me, "You know, the 10th Fleet took a pretty bad licking from the Cardassians in 2374," thus proving that the current U.S. Navy at least has Trekkies, if perhaps not as many geeks as the Air Force.

For its part, the Army's cyber warriors are mostly contained in the Network Enterprise Technology Command, the 9th Signal Command at Fort Huachuca, Arizona. Members of this unit are assigned to the signal commands in each geographic region of the world. Network warfare units, what the Army calls NetWar units, under the Army's Intelligence and Security Command, are also forward-deployed to support combat operations alongside traditional intelligence units. They work closely with NSA to deliver intelligence to war fighters on the ground in Iraq and Afghanistan. The Army Global Network Operations and Security Center, known by the awkward acronym A-GNOSC, manages LandWarNet, which is what the Army calls its portion of the Department of Defense's networks. In July 2008, the Army stood up its first NetWar Battalion. If the Army sounds like the least organized of the services to fight cyber war, that is because it is. After the decision to create Cyber Command was made, the Secretary of Defense mandated the creation of an Army task force to review the Army's cyber mission and organization to support that mission.

While most people who followed the fight over cyber war in the Pentagon thought NSA won it, former NSA Director Ken Minihan was not satisfied, and that gave me pause. Ken is a friend whom I have known since, as an Air Force three-star general, he took over NSA in 1996. He believes that NSA and the U.S. military's approach to cyber operations needs to be rethought. The Navy, he thinks, is focused only on other navies. The Air Force is focused on air defense. The Army is hopelessly lost, and the NSA remains at heart an intelligence collection agency. "Not one of these entities is sufficiently focused on foreign counterintelligence in cyberspace, or on gaining hold of foreign critical infrastructure that the U.S. may want to take down without dropping a bomb in the next conflict." He believes that cyber war planning today lacks a "requirements process," a national-level planning system to get NSA and other organizations working on the same page. "Right now, they are all focused on doing what they want to do, not what a President may need them to be able to do."

Minihan and McConnell are both concerned that U.S. Cyber Command cannot defend the United States. "All the offensive cyber capability the U.S. can muster won't matter if no one is defending the nation from cyber attack," said McConnell. Cyber Command's mission is to defend DoD and maybe some other government agencies, but there are no plans or capabilities for it to defend the civilian infrastructure. Both former NSA Directors believe that mission should be handled by the Department of Homeland Security, as in the existing plans; but both men contend that Homeland has no current ability to defend the corporate cyberspace that makes most of the country work. Neither does the Pentagon. As Minihan put it, "Though it is called the 'Defense' Department, if called on to defend the U.S. homeland from a cyber attack carried out by a foreign power, your half-trillion-dollar-a-year Defense Department would be useless."

THE SECRET ATTEMPT AT A STRATEGY

The perception that cyberspace is a "domain" where fighting takes place, a domain that the U.S. must "dominate," pervades American military thinking on the subject of cyber war. The secret-level National Military Strategy for Cyber Operations (partially declassified as a result of a Freedom of Information Act request) reveals the military's attitude toward cyber war, in part because it was written as a document that we, the citizens, were never supposed to see. It is how they talk about it behind the closed doors of the Pentagon. What is striking in the document is not only the acknowledgment that cyber war is real, but the almost reverential way in which it is discussed as the keystone holding up the edifice of modern war-fighting capability. Because there are so few opportunities to hear from the U.S. military on cyber war strategy, it is worth reading closely the secret-level attempt at a cyber war strategy.

The document, signed out under a cover letter from the Secretary of Defense, declares that the goal is "to ensure the US military [has] strategic superiority in cyberspace." Such superiority is needed to guarantee "freedom of action" for the American military and to "deny the same to our adversaries." To obtain superiority, the U.S. must attack, the strategy declares. "Offensive capabilities in cyberspace [are needed] to gain and maintain the initiative." At first read, the strategy sounds like a mission statement with a bit of zealotry thrown in. On closer examination, however, the strategy reflects an understanding of some of the key problems created by cyber war. Speaking to the geography of cyberspace, the strategy implicitly acknowledges the sovereignty issue ("the lack of geopolitical boundaries . . . allows cyberspace operations to occur nearly anywhere") as well as the presence of civilian targets ("cyberspace reaches across geopolitical boundaries . . . and is tightly integrated into the operations of critical infrastructure and the conduct of commerce"). It does not, however, suggest that such civilian targets should be off-limits from U.S. attacks. When it comes to defending U.S. civilian targets, the strategy passes the buck to the Department of Homeland Security.

The need to take the initiative, to go first, is dictated in part by the fact that actions taken in cyberspace move at a pace never before experienced in war ("cyberspace allows high rates of operational maneuver . . . at speeds that approach the speed of light. . . . [It] affords commanders opportunities to deliver effects at speeds that were previously incomprehensible"). Moreover, the strategy notes that if you do not act quickly, you may not be able to do so because "a previously vulnerable target may be replaced or provided with new defenses with no warning, rendering cyberspace operations less effective." In short, if you wait for the other side to attack you in cyberspace, you may find that the opponent has, simultaneously with their attack, removed your logic bombs or disconnected the targets from the network paths you expected to use to access them. The strategy does not discuss the problems associated with going first or the pressure to do so.

The importance of cyberspace and cyber war to the U.S. military is revealed in the strategy's declaration that "DOD will conduct kinetic missions to preserve freedom of action and strategic

advantage in cyberspace." Translated from Pentagonese, that statement means that rather than cyber attacks being just some support
mechanism of a shooting war, the Defense Department envisions
the need to bomb things in the physical world to defend against
cyber attack, or to drive an enemy into networks that American
cyber warriors control.

The strategic concept of deterrence is discussed in the strategy
only insofar as it envisions a desired end state where "adversaries
are deterred from establishing or employing offensive capabilities
against US interests in cyberspace." Since twenty or thirty nations
have already established offensive cyber units, we apparently did not
deter them from "establishing." The way to stop those nations from
using that capability against us, however, is discussed as "inducing
adversary restraint based on demonstrated capabilities." However,
the secrecy surrounding U.S. offensive cyber war weapons means
that we have no demonstrated capabilities. By the logic of the U.S.
military's strategy, we therefore cannot induce adversary restraint.
The strategy does not suggest a way around this conundrum, let
alone recognize it. Thus, what is called a military strategy for cyber
operations raises some of the key issues that would need to be addressed in a strategy, but it does not provide answers. It is not really
a strategy, but more of an appreciation. To the extent that it provides
guidance, it seems to argue for initiating combat in cyberspace before the other side does, and for doing all that may be needed to
dominate in cyberspace, because to do otherwise would put other
kinds of American dominance at risk.

Buried in the document is, however, a realistic assessment of the
problems facing the U.S. in cyber war: "threat actors can take advantage of [our] dependence" on cyberspace; and, "absent significant
effort, the US will not continue to possess an advantage in cyberspace" and the U.S. will "risk parity with adversaries." Put another
way, the strategy does note the fact that other nations may be able

to inflict cyber war damage on us equal to our ability to inflict it on them. It may actually be worse, because we have a greater dependence on cyberspace, which can play to the advantage of an attacker.

If the U.S. is so vulnerable, to whom is it vulnerable? Who are the other cyber warriors?

WAKE-UP CALL FROM KUWAIT

It may have been the first Gulf War that convinced the generals of China's People's Liberation Army (PLA) that they needed a special advantage, an asymmetrical technical capability against the United States.

It was the first real war the U.S. had fought since Vietnam. In the decades before the 1990–91 Gulf War, the U.S. military had been relatively constrained abroad, by the continued presence of the Soviet Union and its nuclear arsenal. The invasions of Grenada by President Reagan and Panama by the first President Bush had been small engagements in our own backyard, and yet they had not gone terribly well. In those conflicts, U.S. military operations still showed the kind of dysfunction and poor coordination that marked the failed Desert One Mission in Iran in 1979 and helped to end the presidency of Jimmy Carter. Then came Desert Storm. President George H. W. Bush and his cabinet assembled the largest coalition since World War II. More than thirty nations coalesced against Saddam Hussein, bringing together more than 4,000 aircraft, 12,000 tanks, and nearly 2 million military personnel, all paid for by donations from Japan, Germany, Kuwait, and Saudi Arabia. The war was to mark a new era in international relations, what General Brent Scowcroft, President Bush's National Security Advisor, went so far as to call a "new world order." In it, the sovereignty of all nations would be respected and the mission of the United Nations would finally be fulfilled, now that the Soviet Union was no longer in a position to check such actions. Desert Storm was also the dawn of a new kind of warfare, dominated by the computer and other high technology to manage logistics and provide near-realtime intelligence. The Armed Forces Communications and Electronics Association, an American industry group, publicly documented just how dramatically the use of computer networks changed that war in its 1992 book, *The First Information War.*

While General Norman Schwarzkopf and the other military brass may not have been ready to use cyber weapons to take down the Iraqi air defense network, they were ready to embrace computer networks to target the enemy. The war fighters also loved the new breed of "smart weapons" that information systems technology made possible. Designed to replace traditional bombs that required many missions and many tons of munitions dropped to destroy a target, "smart bombs" were designed to put one bomb, and one bomb only, precisely on each target every time. They would greatly reduce the number of missions that needed to be flown and promised to nearly eliminate civilian collateral-damage casualties.

Of course, the "smart weapons" of 1991 were not so smart, and there were not too many of them. In the 1996 movie *Wag the Dog*, a fictional political operative named Conrad "Connie" Brean, played by Robert De Niro, claims that the famous missile down a chimney was done in a studio in Hollywood. "What's the thing people remember about the Gulf War?" Brean asks. "A bomb falling down a chimney. Let me tell you something: I was in the building where we filmed that with a ten-inch model made out of Legos." What De Niro's character claimed wasn't true, but the smart bombs of 1991 were overhyped. While the video was real, the tightly controlled media did not seem to realize that most of the bombs dropped were not precision munitions guided by lasers and satellites but "dumb" bombs, dropped in the thousands by B-52s. The smart bombs then were unreliable and in short supply, but they showed the direction that warfare was moving in, and they showed the Chinese that they were decades behind.

As Desert Storm unfolded, Americans sat glued to their TVs, watching those grainy videos of bombs being dropped down smokestacks. They cheered the renewed prowess of the once-again formidable American military. Saddam Hussein's army was the fourth-largest in the world. His weapons, largely of Soviet make and design, the same as China's arsenal, were mostly destroyed from the air before they could ever be used. The U.S. ground war lasted one hundred hours, following thirty-eight days of air strikes. Among those watching on television were the leaders of the Chinese military. The former Director of National Intelligence, Admiral Mike McConnell, believes that "the Chinese received a big shock when watching the action of Desert Storm." Later they probably read *The First Information War* and other accounts and realized how far behind they really were. They soon began referring to the Gulf War as *zhongda biange*, "the great transformation."

For a period of several years in the mid-1990s the Chinese talked very openly, for a Communist police state, about what they had learned from the Gulf War. They noted that their strategy had been to defeat the U.S. by overwhelming numbers if a war ever happened. Now they concluded that that strategy would not work. They began to downsize their military and invest in new technologies. One of those technologies was *wangluohua*, "networkization," to deal with the "new battlefield of computers." What they talked about publicly then sounds strikingly similar to what the U.S. Air Force generals were saying. Writing in his military's daily paper, one Chinese expert explained that "the enemy country can receive a paralyzing blow through the Internet." A senior colonel, perhaps thinking of the U.S. and China, wrote that "a superior force that loses information dominance will be beaten, while an inferior one that seizes information dominance will be able to win." Major General Wang Pufeng, head of strategy at the military academy, wrote openly of the goal of *zhixinxiquan*, "information dominance." Major General Dai Qingmin of the General Staff stated that such dominance could only be achieved by preemptive cyber attack. These strategists created "Integrated Network Electronic Warfare," something similar to the Netcentric Warfare fad that was sweeping the Pentagon.

By the end of the 1990s, China's strategists had converged on the idea that cyber warfare could be used by China to make up for its qualitative military deficiencies when compared to the United States. Admiral McConnell believes that "the Chinese concluded from the Desert Storm experience that their counter approach had to be to challenge America's control of the battlespace by building capabilities to knock out our satellites and invade our cyber networks. In the name of the defense of China in this new world, the Chinese feel they have to remove that advantage of the U.S. in the event of a war."

A recurring word in these Chinese statements was "asymmetry"; likewise, the phrase "asymmetric warfare." Much of what we know about China's asymmetric warfare doctrine is contained in a slim volume translated as *Unrestricted Warfare*. The book, written by two high-ranking Chinese army colonels, was first published in 1999. It provides a blueprint for how weaker countries can outmaneuver status quo powers using weapons and tactics that fall outside the traditional military spectrum. The publishers of the most widely available English translation view the book as "China's master plan to destroy America," a subtitle the Americans added to the front cover of the U.S. edition. And in case the reader misses the point, the cover shows the World Trade Center engulfed in flames. A quote on the back, from a right-wing lunatic, claims that the book "is evidence linking China to 9-11." Despite the right-wing rhetoric surrounding the U.S. edition, the book is one of the best windows through which we can understand Chinese military thinking on cyber war.

The book advocates tactics that have become known as *shashoujian*, the "assassin's mace," meant to take advantage of weaknesses created by an adversary's seemingly superior conventional capabilities. The goal of the strategy is "fighting the fight that fits one's weapons" and "making the weapons to fit the fight." It proposes a strategy of ignoring the traditional rules of conflict, including, at its extreme, the prohibition on targeting civilians. It also advocates manipulating foreign media, flooding enemy countries with drugs, controlling the markets for natural resources, and joining international legal bodies in order to bend them to one's will. For a book written a decade ago, it also places a heavy emphasis on cyber war.

This possible use of cyber war against a superior force does not mean that China is in fact intent on fighting the U.S., just that its military planners recognize that war with the U.S. is a contingency for which they must plan. The Chinese government has adopted the phrase "peacefully rising" to describe the country's projected emergence as a (if not *the*) global superpower in the twenty-first century. Yet Admiral Mike McConnell believes that "the Chinese are exploiting our systems for information advantage, looking for the characteristics of a weapons system or academic research on plasma physics." China's rapid economic growth and dependence upon global resources, as well as its disputes with its neighbors (Taiwan, Vietnam), probably suggest to its military, however, that they have to be ready for possible conflict someday. And they are getting ready.

To the head of the U.S. military, Admiral Mike Mullen (Chairman of the Joint Chiefs of Staff), it all looks like it is aimed squarely at the United States. "[China is] developing capabilities that are very maritime focused, maritime and air focused, and in many ways, very much focused on us," he said in a speech at the Navy League in May of 2009. "They seem very focused on the United States Navy and our bases that are in that part of the world," he continued. The 2009 update of the annual report from the Office of the Secretary of Defense on the "Military Power of the People's Republic of China" supports these claims. The Chinese have developed long-range radar that can see past our air base on Guam. They have developed antiship missiles that close so fast that none of our defense systems could intercept them. China has purchased one Russian Kuznetsov-class aircraft carrier and is currently in the process of refurbishing it at Dalian shipyard. They will soon have the capability to start constructing new carriers and have put in place a training program so that pilots will be qualified for carrier operations. They have strung over 2,000 missiles along the coast facing Taiwan and are adding more at the rate of 100 per year. They are close to deploying a missile with a 5,000-mile range that could give them a sea-based nuclear strike capability.

It all sounds a bit scary, but look closer and you will see evidence that the modernization alone is insufficient to counter U.S. conventional force superiority. China's military budget is just a fraction of America's. Allegedly only $70 billion, it is less than one-eighth of the Pentagon's budget before adding in the costs of the wars in Afghanistan and Iraq. A U.S. carrier strike group is one of the most powerful conventional forces ever assembled. Consisting of up to a dozen ships, including guided-missile cruisers, destroyers, frigates, submarines, and supply ships, a carrier strike group can cover over 700 nautical miles in a single day, which allows it to go anywhere there is ocean within two weeks. The U.S. Navy boasts eleven carrier battle groups. To keep that force modern, the Navy is in the process of constructing three next-generation Ford-class carriers, with the first carrier set to be launched in 2015.

The Pentagon's annual assessment, *Military Power of the People's Republic of China*, for 2009 estimates that the former Russian aircraft carrier will not be operational before 2015. The consensus view in the U.S. intelligence community is that China is at least a decade away from being able to marshal a modern fighting force that is capable of convincingly defeating even a moderate-sized enemy like Vietnam. Not until 2015 will China be able to project significant power off of its shores, and only then in limited cases against an opponent less capable than the U.S. is now. Unless.

Unless . . . they can even things up by using cyber war against such things as U.S. carriers. The Chinese were always impressed by U.S. carriers, but their attention was heightened in 1996, when President Bill Clinton sent two U.S. carrier battle groups to protect Taiwan during one particularly nasty exchange of tough rhetoric between Beijing and Taipei. So the Chinese military followed its new strategy and developed a "virtual roadmap" for how to take down an aircraft carrier battle group in a paper titled "Tactical Data Links in Information Warfare." This unclassified paper, written by two Chinese Air Force officers, relies on open source material, most of which can be pulled off the web, to illustrate how the information systems that the U.S. military relies on can be jammed or disrupted using relatively low-tech means.

These are the kinds of tactics that *Unrestricted Warfare*'s strategy articulates. The book recommends a program to steal a potential enemy's technology, find flaws in it to exploit, and develop one's own version as part of a program to create a modernized and smaller force. Not lost on Chinese military strategists, however, is the abililty of cyber weapons to skip the battlefield altogether. China has prepared in the event of war to inflict damage on the enemy's home front, not with conventional weapons, but asymmetrically, through cyber attack. The two paths of improvement only make sense together. Even with the significant modernization of equipment, China will not be the equal of the U.S. military for many decades. However, if China can use asymmetrical tactics like cyber war, it believes the new, modern Chinese forces would be sufficiently advanced to take on U.S. forces that will have been crippled by Chinese cyber attack. Recently, Pentagon planners have had a scare put into them by an article in *Orbis* titled "How the United States Lost the Naval War of 2015." In it, James Kraska paints a vivid picture of how in the near future China could take on the United States Navy and win.

THE EAST IS GEEK

From what we know of China's cyber warfare capabilities and the espionage campaigns the Chinese have carried out, that two-pronged approach is exactly what the Chinese have undertaken. Since the late 1990s, China has systematically done all the things a nation would do if it contemplated having an offensive cyber war capability and also thought that it might itself be targeted by cyber war; it has

- created citizen hacker groups,
- engaged in extensive cyber espionage, including of U.S. computer software and hardware,
- taken several steps to defend its own cyberspace,
- established cyber war military units, and
- laced U.S. infrastructure with logic bombs.

While developing cyber strategy, China also made use of private hackers closely aligned with the state's interests. The U.S.-China Economic and Security Review Commission estimates that there are up to 250 groups of hackers in China that are sophisticated enough to pose a threat to U.S. interests in cyberspace. We saw something of their early capabilities in 1999, when the United States led a NATO air campaign to stop the slaughter in Kosovo by Serbian forces. The U.S. had all but perfected its smart weapons and used them to eliminate the Serbians' Soviet-era military apparatus without losing a single American life (one U.S. warplane went down due to mechanical failure). Unfortunately, smart weapons can't make up for bad intelligence. Six bombs dropped from U.S. aircraft hit the precise coordinates provided to the mission planners by the CIA. The target was supposed to be the Yugoslav Federal Directorate for Supply and Procurement, a planning agency of the Serbian military. The coordinates, however, were about 900 feet off from the Directorate and exactly on top of the Chinese embassy.

The Chinese held protests outside U.S. embassies and consulates, issued condemnatory statements within the UN and other bodies, and demanded compensation for the victims and their families. After the embassy bombing, U.S. and NATO websites were targeted with denial of service attacks. Government agencies had their inboxes stuffed with spam messages protesting the bombing. Some NATO webpages were forced down, while others were defaced. The attacks did little damage to U.S. military or government operations. The effort amounted to little more than what we call "hacktivism" today, a fairly mild form of online protest. It was, however, a first use of cyberspace by China to protest. Chinese hacktivists did it again in 2001, when a U.S. "spy plane" allegedly entered Chinese airspace and was forced by Chinese fighter jets to land in China. However, while these Chinese citizen hackers were launching their primitive denial of service and spam attacks, China's intelligence-industry partnership was also busy.

The Chinese government went after two underpinnings of the U.S. computer industry's dominance of networking technology, Microsoft and Cisco. By threatening to ban Chinese government procurement from Microsoft, Beijing persuaded Bill Gates to provide China with a copy of its secret operating system code. Microsoft had refused to show that same code to its largest U.S. commercial customers. Then China copied the Cisco network router found on almost all U.S. networks and at most Internet service providers.

Cisco had a manufacturing plant for the routers in China. Chinese companies then sold counterfeit Cisco routers at cut-rate discounts around the world. The buyers allegedly included the Pentagon and other federal government entities. Counterfeit routers started showing up on the market in 2004. Three years later, the FBI and the Justice Department indicted two brothers who owned a company called Syren Technology for selling the counterfeit routers to a customer list that included the Marine Corps, the Air Force, and multiple defense contractors. A fifty-page report authored by the FBI and circulated within the technology industry concluded that the routers could be used by foreign intelligence agencies to take down networks and "weaken cryptographic systems." Meanwhile, another Chinese company, Huawei, was selling similar routers throughout Europe and Asia. The major difference was that, unlike the counterfeits, these routers did not say Cisco on the front. Their label said Huawei.

With intimate knowledge of the flaws in Microsoft and Cisco software and hardware, China's hackers could stop most networks from operating. But wouldn't the Chinese be vulnerable, too? They would be, if they used the same Microsoft and Cisco products we do. As part of the deal with Microsoft, the Chinese modified the version sold in their country to introduce a secure component using their own encryption. Hedging their bets, they also developed their own operating system, called Kylin, modeled on the stable open source system known as Free BSD. Kylin was approved by the People's Liberation Army for use on their systems. China allegedly also developed its own secure microprocessor for use on servers and Huawei routers. The Chinese government is trying to install "Green Dam Youth Escort" software on all of its computers, allegedly to screen for child pornography and other prohibited material. If they get it to work, and proliferate it on all their systems, Green Dam could also scan for malware installed by enemy states.

In addition to Green Dam, there is the system that U.S. wags call the Great Firewall of China. Not really a firewall, the government-run system screens traffic on ISPs for subversive material, such as the Universal Declaration of Human Rights. The system engages in something called "Domain Name System hijacking," sending you to a Chinese government clone of a real site when you are in China and try to go, for example, to the webpage of a Christian evangelical organization. It also has the ability to disconnect all Chinese networks from the rest of the global Internet, something that would be handy to have if you thought the U.S. was about to launch a cyber war attack on you. James Mulvenon, one of the leading American experts on China's cyber war capabilities, says that taken together, Green Dam, the Great Firewall, and other systems represent "a substantial investment by Chinese authorities in enhanced blocking, filtering, and monitoring" of their own cyberspace.

By 2003, China had announced the creation of cyber warfare units. Housed at the naval base on Hainan Island are the Third Technical Department of the PLA and the Lingshui Signals Intelligence Facility. According to the Pentagon, these units are responsible for offense and defense in cyberspace, and have designed cyber weapons that have never been seen before and that no defenses have been designed to stop. In one publication, the Chinese listed ten examples of such weapons and techniques:

- planting information mines
- conducting information reconnaissance
- changing network data
- releasing information bombs
- dumping information garbage
- disseminating propaganda
- applying information deception
- releasing clone (*sic*) information

- organizing information defense
- establishing network spy stations

China did establish two "network spy stations," not far from the U.S., in Cuba. With the permission of the Castro government, the Chinese military created a facility to monitor U.S. Internet traffic and another to monitor DoD communications. At about the same time China announced the creation of its cyber warfare units, the U.S. experienced one of the worst episodes of cyber espionage to date. Known as Titan Rain, the U.S. code name given to the case, the incident involved the extraction of between 10 and 20 terabytes of data off the Pentagon's unclassified network. The hackers also targeted the defense contractor Lockheed Martin, other military sites, and, for reasons that remain hard to fathom, the World Bank. Vulnerabilities in Pentagon and other targeted networks were systematically identified and then exploited to extract information through servers in South Korea and Hong Kong. Investigators were able to trace the flow from these intermediate servers back to a final server in Guangdong, China. U.S. Air Force Major General William Lord directly and publicly attributed the attacks not to Chinese hacktivists, but to the Chinese *government*.

By 2007, the Chinese government seemed to be involved in a widespread series of penetrations of U.S. and European networks, successfully copying and exporting huge volumes of data. The Director of the British domestic intelligence service MI5, Jonathan Evans, wrote letters to 300 leading companies in the U.K., advising them that their networks had probably been penetrated by the Chinese *government*. Evans's counterpart in Germany, Hans Remberg, also accused the Beijing government, this time of hacking into the computer of Angela Merkel, the German Chancellor.

The computer espionage also went after a high-ranking American, hacking into the computer of Secretary of Defense Robert

Gates. Later, Chinese operatives copied information off of U.S. Secretary of Commerce Carlos Gutierrez's laptop when he visited Beijing, then attempted to use that information to gain access to Commerce Department computers. Commenting on the Chinese, Gates's Deputy Undersecretary, Robert Lawless, admitted that they have "a very sophisticated capability to attack and degrade our computer systems . . . to shut down our critical systems. They see it as a major component of their asymmetrical warfare capability."

In 2009, Canadian researchers uncovered a highly sophisticated computer program they dubbed GhostNet. It had taken over an estimated 1,300 computers at several countries' embassies around the world. The program had the capability to remotely turn on a computer's camera and microphone without alerting the user and to export the images and sound silently back to servers in China. A top target of the program were offices related to nongovernmental organizations working on Tibetan issues. The operation ran for twenty-two months until discovered. The same year, U.S. intelligence leaked to the media that Chinese hackers had penetrated the U.S. power grid and left behind tools that could be used to bring the grid down.

The extent of Chinese government hacking against U.S., European, and Japanese industries and research facilities is without precedent in the history of espionage. Exabytes of data have been copied from universities, industrial labs, and government facilities. The secrets behind everything from pharmaceutical formulas to bioengineering designs, to nanotechnology, to weapons systems, to everyday industrial products have been taken by the People's Liberation Army and by private hacking groups and given to China, Inc.

In the latest incident to become public, Google revealed its discovery of a highly sophisticated campaign targeting both the company's intellectual property and the e-mail accounts of leaders in the Chinese dissident movement.

The hackers used advance "spear-phishing" techniques to dupe senior Google executives into visiting websites where malware would automatically be downloaded onto their computers to give the hackers root access. While most phishing scams cast a wide net and try to catch a few peope who are gullible enough to fall for Nigerian scammer e-mails, spear-phishing specifically targets an individual, figures out who their acquaintances are on Facebook or Linked-in, and then tailors a message to look like it is from someone they would trust. If you were a senior research scientist at Google, you might have received an e-mail containing a link to a website that looked like it was from a colleague. The message might have said, "Hey, Chuck, I think this story will interest you . . ." and then provided a link to fairly innocuous site. When the target clicked on the link and visited the site, the hackers used a zero-day flaw in Internet Explorer, one that was not publicly known and had yet to be patched, to download the malware silently and in such a fashion that no antivirus software or other measures would detect it. The malware created a back door to the computer so the hackers could maintain their access and used the first compromised computer to work their way across the corporate network until they reached the servers containing the source code, the crown jewel of a software company.

When Google's scientists figured out what was going on in mid-December, they traced back the hacking to a server in Taiwan, where they found copies of their proprietary information and those of at least twenty other companies, including Adobe, Dow Chemical, and the defense contractor Northrop Grumman. From there, they traced the attacks back to Mainland China, and then went to the FBI, making their public announcement of the hacking and plans to exit the Chinese market in mid-January.

Some will suggest that war with China is, in any event, unlikely.

China's dependence on U.S. markets for its manufactured goods and the trillions the country has invested in U.S. Treasury bills mean that China would have a lot to lose in a war. One Pentagon official who spoke on the condition of anonymity isn't so sure. He points out that the economic meltdown in the U.S. has had a secondary effect in China that has put millions of Chinese factory workers out on the streets. The Chinese government has not shown the kind of concern that we expect in the West and is not apparently worried about any weakening of its grip on the Chinese people. The lesson the Pentagon official takes away is that China can take economic lumps and may well do so if the gains from warfare are perceived as high enough.

What might such gains be? The trite answer one often hears is that China may find itself forced to stop Taiwan from implementing a declaration of independence. When serious analysts weigh the prospects of open conflict with China, however, they see it playing out over the open waters of the South China Sea. The Spratly Islands are not exactly a tourist destination. They are not exactly islands. If all were piled up together, the reefs, sandbars, and rocks in the South China Sea would amount to less than two square miles of land. That two square miles of land is spread out over more than 150,000 square miles of ocean. It's not the islands that China, Vietnam, Taiwan, Malaysia, the Philippines, and Brunei are feuding over, but what is under them and around them. The reefs have some of the largest remaining stocks of fish in the world, a resource not to be discounted among the growing and hungry nations that lay claim to the waters. The islands also skirt the critical trade route that links the Indian Ocean to the Pacific nations through which a large majority of the world's oil flows out of the Middle East. Then there are the Spratlys' oil and gas. Undeveloped fields estimated to hold more natural gas than are Kuwait, currently home to the fourth-largest reserves in the world, could fuel the economies of any of the countries for decades to come. Oil fields in the islands are already well developed, often with platforms established by several nations drawing out of the same reservoir.

If China decides to flex its newly developed military muscle, it may very well be in an attempt to wrest these islands from its neighbors, a scenario explored as a tabletop exercise later in the book. If China does seize the islands, the U.S. could, though reluctantly, be drawn into a response. The U.S. has established security guarantees with both the Philippines and Taiwan. Chevron has helped Vietnam develop the offshore oil fields that that nation claims.

Alternatively, we might be deterred from intervening against China in the Pacific Rim if the costs of doing so would be significant damage or disruption at home. According to Defense Secretary Robert Gates, cyber attacks "could threaten the United States' primary means to project its power and help its allies in the Pacific." Is that enough to deter the U.S. from a confrontation with China? If the possibility of China crippling our force projection capability is not enough to deter us, maybe the realization of our domestic vulnerabilities to cyber attack would be. The alleged emplacement of logic bombs in our electric grid may have been done in such a way that we would notice. One former government official told us that he suspects the Chinese wanted us to know that if we intervened in a Chinese conflict with Taiwan, the U.S. power grid would likely collapse. "They want to deter the United States from getting involved militarily within their sphere of influence."

The problem is, however, that deterrence only works if the other side is listening. U.S. leaders may not have heard, or fully understood, what Beijing was trying to say. The U.S. has done little or nothing to fix the vulnerabilities in its power grid or in other civilian networks.

A SCORE OF OTHERS

I focused on China because its cyber war development has been, oddly, somewhat transparent. U.S. intelligence officials do not, however, rate China as the biggest threat to the U.S. in cyberspace. "The Russians are definitely better, almost as good as we are," said one. There seems to be a consensus that China gets more attention because, intentionally or otherwise, it has often left a trail of bread crumbs that can be followed back to Tiananmen Square.

The Russian nongovernmental hackers, including large cyber criminal enterprises, are a real force in cyberspace, as was demonstrated in the attacks on Estonia and Georgia discussed in chapter 1. The hacktivists and criminals are generally thought to be sanctioned by what used to be called the Sixteenth Directorate, a part of the infamous Soviet intelligence apparatus known as the KGB. Later it was called FAPSI. Few American intelligence officers could ever remember what FAPSI stood for (it's the Russian acronym for: Federal Commission for Government Communications and Information), they just knew it was "Moscow's NSA."

Like America's NSA, FAPSI started out doing code making and breaking, radio intercept, bugging, and wiretapping. As soon as the Internet appeared, however, FAPSI was on to it, taking over the largest ISP in Russia, later requiring all Russian ISPs to install monitoring systems that only FAPSI could access. Of course, during the rise of the Internet, the Soviet Union ended, and so, theoretically, did the KGB and FAPSI. In fact, the organizations merely put up their headquarters with new names. After several changes, in 2003 FAPSI became the Service of Special Communications and Information. Not all of their placarded buildings are in Moscow. In the southern city of Voronezh, FAPSI, as many Russians still call it, runs what might be the largest (and certainly one of the best) hacker schools in the world. By now, of course, they are probably calling themselves cyber warriors.

Other nations known to have skilled cyber war units are Israel and France. U.S. intelligence officials have suggested that there are twenty to thirty militaries with respectable cyber war capability, including those of Taiwan, Iran, Australia, South Korea, India, Pakistan, and several NATO states. "The vast majority of the industrialized countries in the world today have cyber-attack capabilities," said former Director of National Intelligence Admiral Mike McConnell.

WHEN CYBER WARRIORS ATTACK

You may by now believe that there are cyber warriors, but in addition to jamming Internet sites what can they do, really? Obviously, we have not had a full-scale cyber war yet, but we have a good idea what it would look like if we were on the receiving end. Imagine a day in the near future. You are the Assistant to the President for Homeland Security and you get a call from the White House Situation Room as you are packing up to leave the office for the day, at eight p.m. NSA has issued a "CRITIC" message, a rare alert that something important has just happened. The one-line message says only: "large scale movement of several different zero day malware programs moving on Internet in the US, affecting critical infrastructure." The Situation Room's Senior Duty Officer suggests that you come down and help him figure out what is going on.

By the time you get to the Situation Room, the Director of the Defense Information Systems Agency is waiting on the secure phone for you. He has just briefed the Secretary of Defense, who suggested he call you. The unclassified Department of Defense network known as the NIPRNET is collapsing. Large-scale routers throughout the network are failing, and constantly rebooting. Network traffic is essentially halted. As he is telling you this, you can hear someone in the background trying to get his attention. When the general comes back on the line, he says softly and without emotion, "Now it's happening on the SIPRNET and JWICS, too." He means that DoD's classified networks are grinding to a halt.

Unaware of what is happening across the river at the Pentagon, the Undersecretary of Homeland Security has called the White House, urgently needing to speak to you. FEMA, the Federal Emergency Management Agency, has told him that two of its regional offices, in Philadelphia and in Denton, Texas, have reported large refinery fires and explosions in Philadelphia and Houston, as well as lethal clouds of chlorine gas being released from several chemical plants in New Jersey and Delaware. He adds that the U.S. Computer Emergency Response Team in Pittsburgh is being deluged with reports of systems failing, but he hasn't had time to get the details yet.

Before you can ask the Senior Duty Officer where the President is, another officer thrusts a phone at you. It's the Deputy Secretary of Transportation. "Are we under attack?" she asks. When you ask why, she ticks off what has happened. The Federal Aviation Administration's National Air Traffic Control Center in Herndon, Virginia, has experienced a total collapse of its systems. The alternate center in Leesburg is in a complete panic because it and several other regional centers cannot see what aircraft are aloft and are trying to manually identify and separate hundreds of aircraft. Brickyard, the Indianapolis Center, has already reported a midair collision of two 737s. "I thought it was just an FAA crisis, but then the train wrecks started happening . . ." she explains. The Federal Railroad Administration has been told of major freight derailments in Long Beach, Norfolk, Chicago, and Kansas City.

Looking at the status board for the location of the President, you see it says only "Washington-OTR." He is on an "off the record," or personal, activity outside the White House. Reading your mind,

the Senior Duty Officer explains that the President has taken the
First Lady to a hip new restaurant in Georgetown. "Then put me
through to the head of his Secret Service detail," says a breathless
voice. It's the Secretary of the Treasury, who has run from his office
in the building next to the White House. "The Chairman of the Fed
just called. Their data centers and their backups have had some sort
of major disaster. They have lost all their data. Its affecting the data
centers at DTCC and SIAC—they're going down, too." He explains
that those initials represent important financial computer centers in
New York. "Nobody will know who owns what. The entire financial
system will dissolve by morning."

As he says that, your eyes are drawn to a television screen reporting on a derailment on the Washington Metro in a tunnel under the
Potomac. Another screen shows a raging flame in the Virginia sub-
urbs where a major gas pipeline has exploded. Then the lights in the
Situation Room flicker. Then they go out. Battery-operated emergency spotlights come on, casting the room in shadows and bright
light. The television flat screens and the computer monitors have
gone blank. The lights flicker again and come back on, as do some
of the screens. There is a distant, loud droning. "It's the backup
generator, sir," the Duty Officer says. His deputy again hands you a
secure phone and mouths the words you did not want to hear: "It's
for you. It's POTUS."

The President is in the Beast, his giant armored vehicle that re-
sembles a Cadillac on steroids, on his way back from the restaurant.
The Secret Service pulled him out of the restaurant when the blackout hit, but they are having a hard time getting through the traffic.
Washington's streets are filled with car wrecks because the signal
lights are all out. POTUS wants to know if it's true what his Secret
Service agent told him, that the blackout is covering the entire eastern half of the country. "No, wait, what? Now they're saying that

the Vice President's detail says it's out where he is, too. Isn't he in San Francisco today? What time is it there?"

You look at your watch. It's now 8:15 p.m. Within a quarter of an hour, 157 major metropolitan areas have been thrown into knots by a nationwide power blackout hitting during rush hour. Poison gas clouds are wafting toward Wilmington and Houston. Refineries are burning up oil supplies in several cities. Subways have crashed in New York, Oakland, Washington, and Los Angeles. Freight trains have derailed outside major junctions and marshaling yards on four major railroads. Aircraft are literally falling out of the sky as a result of midair collisions across the country. Pipelines carrying natural gas to the Northeast have exploded, leaving millions in the cold. The financial system has also frozen solid because of terabytes of information at data centers being wiped out. Weather, navigation, and communications satellites are spinning out of their orbits into space. And the U.S. military is a series of isolated units, struggling to communicate with each other.

Several thousand Americans have already died, multiples of that number are injured and trying to get to hospitals. There is more going on, but the people who should be reporting to you can't get through. In the days ahead, cities will run out of food because of the train-system failures and the jumbling of data at trucking and distribution centers. Power will not come back up because nuclear plants have gone into secure lockdown and many conventional plants have had their generators permanently damaged. High-tension transmission lines on several key routes have caught fire and melted. Unable to get cash from ATMs or bank branches, some Americans will begin to loot stores. Police and emergency services will be overwhelmed.

In all the wars America has fought, no nation has ever done this kind of damage to our cities. A sophisticated cyber war attack by one of several nation-states could do that today, in fifteen minutes, without a single terrorist or soldier ever appearing in this country. Why haven't they done it by now, if they can? For the same reason that the nine nations with nuclear weapons haven't used one of them since 1945, because they need to have the political circumstances that cause them to believe such an attack would be in their interest. But unlike with nuclear weapons, where an attacker may be deterred by the promise of retaliation or by the radioactive blowback on his own country, launching a cyber attack may run fewer risks. In cyber war, we may never even know what hit us. Indeed, it may give little solace to Americans shivering without power to know that the United States may be about to retaliate in kind.

"While you were on the line with the President, sir, Cyber Command called from Fort Meade. They think the attack came from Russia and they are ready to turn out the lights in Moscow, sir. Or maybe it was China, so they are ready to hit Beijing, if you want to do that. Sir?"

CHAPTER THREE

THE BATTLESPACE

Cyberspace. It sounds like another dimension, perhaps with green lighting and columns of numbers and symbols flashing in midair, as in the movie *The Matrix*. Cyberspace is actually much more mundane. It's the laptop you or your kid carries to school, the desktop computer at work. It's a drab windowless building downtown and a pipe under the street. It's everywhere, everywhere there's a computer, or a processor, or a cable connecting to one.

And now it's a war zone, where many of the decisive battles in the twenty-first century will play out. To understand why, we need to answer some prior questions, like: What is cyberspace? How does it work? How can militaries fight in it?

HOW AND WHY CYBER WAR IS POSSIBLE

Cyberspace is all of the computer networks in the world and everything they connect and control. It's not just the Internet. Let's be
clear about the difference. The Internet is an open network of networks. From any network on the Internet, you should be able to
communicate with any computer connected to any of the Internet's
networks. Cyberspace includes the Internet *plus* lots of other networks of computers that are not supposed to be accessible from the
Internet. Some of those private networks look just like the Internet,
but they are, theoretically at least, separate. Other parts of cyberspace are transactional networks that do things like send data about
money flows, stock market trades, and credit card transactions.
Some networks are control systems that just allow machines to speak
to other machines, like control panels talking to pumps, elevators,
and generators.

What makes these networks a place where militaries can fight?
In the broadest terms, cyber warriors can get into these networks
and control or crash them. If they take over a network, cyber warriors could steal all of its information or send out instructions that
move money, spill oil, vent gas, blow up generators, derail trains,
crash airplanes, send a platoon into an ambush, or cause a missile
to detonate in the wrong place. If cyber warriors crash networks,
wipe out data, and turn computers into doorstops, then a financial
system could collapse, a supply chain could halt, a satellite could
spin out of orbit into space, an airline could be grounded. These are
not hypotheticals. Things like this have already happened, sometimes experimentally, sometimes by mistake, and sometimes as a
result of cyber crime or cyber war. As Admiral Mike McConnell
has noted, "information managed by computer networks—which
run our utilities, our transportation, our banking and communi-

cations—can be exploited or attacked in seconds from a remote location overseas. No flotilla of ships or intercontinental missiles or standing armies can defend against such remote attacks located not only well beyond our borders, but beyond physical space, in the digital ether of cyberspace."

Why, then, do we run sophisticated computer networks that allow unauthorized access or unauthorized commands? Aren't there security measures? The design of computer networks, the software and hardware that make them work, and the way in which they were architected, create thousands of ways that cyber warriors can get around security defenses. People write software and people make mistakes, or get sloppy, and that creates opportunities. Networks that aren't supposed to be connected to the public Internet very often actually are, sometimes without their owners even knowing. Let's look at some things in your daily life as a way of explaining how cyber war can happen. Do you think your condominium association knows that the elevator in your building is, like *ET* in the movie of the same name, "phoning home"? Your elevator is talking over the Internet to the people who made it. Did you know that the photocopier in your office is probably doing the same thing? Julia Roberts's character in the recent movie *Duplicity* knew that many copying machines are connected to the Internet and can be hacked, but most people don't know that their copier could even be online. Even fewer think about the latest trick, shredders that image. Just before all those sensitive documents pass through the knives that cut them into little pieces, they go by a camera that photographs them. Later, the cleaning crew guy will take his new collection of pictures away to whoever hired him.

Your elevator and copier "phoning home" is supposed to be happening, the software is working properly. But what if your competitor has a computer programmer who wrote a few lines of code and slipped them into the processor that runs your photocopier? Let's say those few lines of computer code instruct the copier to store an image of everything it copies and put them into a compressed data (or zip) file. Then, once a day, the copier accesses the Internet and—ping!—it shoots that zip file across the country to your competitor. Even worse, on the day before your company has to submit a competitive bid for a big contract—ping!—the photocopier catches fire, causing the sprinklers to turn on, the office to get soaked, and your company to be unable to get its bid done in time. The competitor wins, you lose.

Using an Internet connection you did not know existed, someone wrote software and downloaded it onto your photocopier, which you did not even know had an onboard processor big enough to be a computer. Then that someone used the software to make the photocopier do something it wouldn't otherwise do, short-circuit or jam and overheat. They knew the result would be a fire. They probably experimented with a copier just like yours. The result is your office is flooded by the sprinkler system and you think it was an accident. Somebody reached out from cyberspace and made your physical space a mess. That someone is a hacker. Originally "hacker" meant just somebody who could write instructions in the code that is the language of computers to get them to do new things. When they do something like going where they are not authorized, hackers become cyber criminals. When they work for the U.S. military, we call them cyber warriors.

In this scenario, the cyber criminal used the Internet as his avenue of attack, first to get information and then to do damage. His weapon was a few lines of software, which he inserted into the computer in the copier. Or you may think about it this way: he used software to turn your copier into the weapon. He succeeded because the software program that ran the copier was written to allow people to add commands and give those commands remotely. The designers of the copier never thought anyone would make it a weapon, so they never wrote their software in a manner that would make that difficult or impossible to do. The same is true of the designers of the electric power grid and other systems. They didn't think about people hacking them and turning their systems into weapons. Your office manager didn't pay attention when the salesperson said the copier would have a remote diagnostics capability to download improvements, fix problems, and dispatch a repairman with the right replacement parts. Hackers paid attention, or maybe they were just exploring their cyberspace neighborhood and found an address that identified itself as "Xeonera Copier 2000, serial number 20-003488, at Your Company, Inc."

If you doubt that copiers are part of cyberspace, read *Image Source Magazine*:

> Historically, remote diagnostics required dial up modems. The methodology at that time was somewhat of an inconvenience to the customer and very expensive for the dealer who had to install phone jacks near each device and switch boxes to be compatible with their client's phone systems. But those barriers have now been eliminated with the introduction of the web and wireless networks. Now that all networked devices have an address, a diagnostic error report can be transmitted in real time via the web and technicians can be dispatched by the device itself, sometimes before the customer knows there is a problem. Today, there is no excuse for any service organization to ignore the cost savings and value of remote diagnostics. Virtually every printer manufacturer either has their own remote diagnostic tool (i.e. Ricoh's Remote, Kyocera Admin, Sharp's Admin, Xerox's DRM) or have partnered with third party companies like Imaging Portals or Print Fleet.

While mundane, this hypothetical scenario is helpful because it shows the three things involved in cyberspace that make cyber war possible: (1) flaws in the design of the Internet; (2) flaws in hardware and software; and (3) the move to put more and more critical systems online. Let's look at each.

VULNERABILITIES OF THE INTERNET

There are at least five major vulnerabilities in the design of the Internet itself. The first of these is the addressing system that finds out where to go on the Internet for a specific address.

ISPs are sometimes called "carriers," because they are the companies that carry the Internet's traffic. Other companies make the computer terminals, the routers, the servers, the software, but it is the ISPs that link them all together. All ISPs are not created equal. For our discussion, let's divide them into two categories. There are the national ISPs that own and operate thousands of miles of fiber-optic cable running from coast to coast, connecting all the big cities. There are six of these big ISPs in the United States (Verizon, AT&T, Qwest, Sprint, Level 3, and Global Crossing). Because their big fiber-optic cable pipes form the spine of the Internet in the U.S., they are called the "backbone providers," or, more technically, the Tier 1 ISPs. Once they get the backbone into your city, they connect up with lots of smaller ISPs that run service to local businesses and to your house. Your local ISP is probably the phone company or the cable TV company. (If it's the phone company, it may be that you have one of the Tier 1 ISPs also providing your local service.) Their wires run from your house down the street to the world.

To see how this works, and to discover some of the vulnerabilities of the Internet addressing system, follow what happens when I connect to the Internet. I open a "browser" on my laptop. Just by opening the browser, I am requesting that it go out onto the Internet and bring back "my homepage." Let's say that "homepage" is that of the consulting firm where I work. So, sitting in my home office in

Rappahannock County, Virginia, in the foothills of the Blue Ridge Mountains, I click and my browser goes to www.mycompany.com. Since computers can't understand words like "mycompany," the address needs to be translated into 1's and 0's that computers can read. To do that, my browser uses the Domain Name System. Think of it as the 411 information operator. You say a name, you get a number.

My consulting firm is headquartered seventy-five miles away from my home in Virginia, but its webpage is hosted on a remote server in Minneapolis with the Internet address of, let's say, 123.45.678.90. That is a lot of numbers to memorize. Luckily I don't have to. The browser uses the Domain Name System to look up the address. The browser sends a message to a database kept on a server computer, part of an elaborate hierarchy of such computers that together form the Domain Name System. For cyber warriors, the Domain Name System is a target. It was designed with little thought to security, so hackers can change its information and misdirect you to a phony webpage.

When I open up the browser it sends a request to the server hosting the page. The request is broken down into a series of packets that are each sent individually. Let's trace just one packet along its way from my computer to the website. The first hop is from my computer to the wi-fi card in my computer, where the packets are translated into radio waves and sent out over the air to the wi-fi router in my house. If that router is improperly secured, hackers can get into the computer over the wi-fi connection. The wi-fi router turns the signal back from a radio wave into an electronic signal passed to my local ISP in the booming megalopolis of Culpepper, Virginia.

If you know it, you may think that Culpepper is lovely, but not necessarily near the heart of cyberspace. Because it's just beyond the blast radius if a nuclear weapon were to go off in Washington, the government and the financial community have all sorts of databases nearby. So, there is an AT&T node there, at 13456 Lovers Lane.

(Really.) My ISP has a line running across town to the AT&T facility, where the electrons of my request for the webpage get converted into photons so they can hop on AT&T's fiber-optic network. Once on the fiber, the packet first hits a router in Morristown, New Jersey, is passed to another AT&T router in Washington, D.C., and then back to New Jersey, this time to a router in Middletown.

At Middletown, the router passes the packet to another Tier 1 company, Level 3. Once on the Level 3 backbone, the packet is routed through three different nodes in Washington, D.C. At this point, the packet has traveled over radio waves, copper wires, and high-speed bundles of fiber-optic cables for more than 800 miles, but is only about 75 miles from where I first sent it off. The last Level 3 router, in Washington, sends it speeding toward Chicago (now we are getting somewhere), where it descends through two more Level 3 routers before being sent to Minneapolis. What goes to Minneapolis, though, does not necessarily stay in Minneapolis. Instead of handing off to our web hosting provider, the packet goes another 741 miles to another Level 3 router in the company's headquarters in Broomfield, Colorado, which then routes the packet back to our company's ISP, in Minneapolis, and on to our web server. To travel the 900 miles to Minneapolis, the packet went about 2,000 miles out of its way, but the whole process took a few seconds. It also provided several opportunities for cyber warriors.

If cyber warriors had wanted to send those packets to the wrong place, or to prevent them from going anywhere, they had at least two opportunities. First, as noted earlier, they could have attacked that Internet 411, the Domain Name System, and sent me to the wrong page, perhaps to a phony look-alike webpage, where I would enter my account number and password. Rather than hacking the Domain Name System to hijack a webpage request, however, cyber warriors could attack the system itself. This is just what happened in February 2007, when six of the thirteen top-level worldwide domain servers were targeted in a DDOS attack. Similar to the botnets that hit Estonia and Georgia, the attack flooded the domain name servers with thousands of requests per second. Two of the servers targeted were taken down, including one that handles traffic for the Department of Defense. The other four were able to manage the attack by shifting requests to other servers not targeted in the attack. The attack was traced back to the Pacific region and lasted only eight hours. The attackers stopped it either because they were afraid continuing it would allow investigators to trace it back to them or, more likely, because they were just testing to see if they could do it.

In 2008, the hacker Dan Kaminsky showed how a sophisticated adversary could hack the system. Kaminsky released a software tool that could quietly access the Domain Name System computers and corrupt the database of name addresses and their related numbered addresses. The system would then literally give you a wrong number. Just misdirecting traffic could cause havoc with the Internet. One cyber security company found twenty-five different ways it could hack the Domain Name System to cause disruption or data theft.

The second vulnerability of the Internet is routing among ISPs, a system known as the Border Gateway Protocol. Another opportunity for a cyber warrior in the one-second, 2,000-mile trip of packets from my home came when they jumped onto the AT&T network. AT&T runs the most secure and reliable Internet service in the world, but it is as vulnerable as anyone else to the way the Internet works. When the packets got on the backbone, they found that AT&T does not connect directly to my company. So who does? The packets checked a database that all of the major ISPs contribute to. There they found a posting from Level 3 that said, in effect, "If you want to connect to mycompany.com, come to us." This routing system regulates traffic at the points where the ISPs come together, where one starts and the other stops, at their borders.

BGP is the main system used to route packets across the Internet.

The packets have labels with a "to" and "from" address, and BGP is the postal worker that decides what sorting station the packet goes to next. BGP also does the job of establishing "peer" relationships between two routers on two different networks. To go from AT&T to Level 3 requires that an AT&T router and a Level 3 router have a BGP connection. To quote from a report from Internet Society, a nonprofit organization dedicated to developing Internet-related standards and policies, "There are no mechanisms internal to BGP that protect against attacks that modify, delete, forge, or replay data, any of which has the potential to disrupt overall network routing behavior." What that means is that when Level 3 said, "If you want to get to mycompany.com, come to me," nobody checked to see if that was an authentic message. The BGP system works on trust, not, to borrow Ronald Reagan's favorite phrase, on "trust but verify." If a rogue insider working for one of the big ISPs wanted to cause the Internet to seize up, he could do it by hacking into the BGP tables. Or someone could hack in from outside. If you spoof enough BGP instructions, Internet traffic will get lost and not reach its destination.

Everyone involved in network management for the big ISPs knows about the vulnerabilities of the Domain Name System and the BGP. People like Steve Kent of BBN Labs in Cambridge, Massachusetts, have even developed ways of eliminating those vulnerabilities, but the Federal Communications Commission has not required the ISPs to do so. Parts of the U.S. government are deploying a secure Domain Name System, but the practice is almost nonexistent in the commercial infrastructure. Decisions on the Domain Name System are made by a nongovernmental international organization called ICANN (pronounced "*eye*-can"), which is unable ("*eye-cannot*") to get agreement on a secure system. The result is that the Internet itself could easily be a target for cyber warriors, but most cyber security experts think that unlikely because the Internet is so useful for attacking other things.

ICANN demonstrates the second vulnerability of the Internet, which is governance, or lack thereof. No one is really in charge. In the early days of the Internet, ARPA (DoD's Advanced Research Project Agency) filled the function of network administrator, but nobody does now. There are technical bodies, but few authorities. ICANN, the Internet Corporation for Assigned Names and Numbers, is the closest that any organization has come to being responsible for the management of even one part of the Internet system. ICANN ensures that web addresses are unique. Computers are logical devices, and they don't deal well with ambiguity. If there were two different computers on the Internet each with the same address, routers would not know what to do. ICANN solves that problem by working internationally to assign addresses. ICANN solves one of the problems of Internet governance, but not a host of other issues. More than a dozen intergovernmental and nongovernmental organizations play some role in Internet governance, but no authority provides overall administrative guidance or control.

The third vulnerability of the Internet is the fact that almost everything that makes it work is open, unencrypted. When you are crawling around the web, most of the information is sent "in the clear," meaning that it is unencrypted. It's like your local FM classic rock station broadcasting Pink Floyd and Def Leppard "in the clear" so that anyone tuned to that channel can receive the signal and rock along rolling down the highway. A radio scanner purchased at Radio Shack can listen in on the two-way communications between truckers, and in most cities, between police personnel. In some cities, however, the police will "scramble" the signal so that criminal gangs cannot monitor police comms. Only someone with a radio that can unencrypt the traffic can hear what is being said. To everyone else, it just sounds like static.

The Internet generally works the same way. Most communication is openly broadcast, and only a fraction of the traffic is encrypted.

The only difference is that it is a little more difficult to tune in to someone else's Internet traffic. ISPs have access (and can give it to the government), and mail-service providers like Google's Gmail have access (even if they say they don't). In both those cases, by using their services you are more or less agreeing that they may be able to see your web traffic or e-mails. For a third party to get access, they need to do what security folks call "snoop" and use a "packet sniffer" to pick up the traffic. A packet sniffer is basically a wiretap device for Internet traffic and can be installed on any operating system and used to steal other people's traffic on a local area network. When plugged into a local or an Ethernet network, any user on the system can use a sniffer to pull in all the other traffic. The standard Ethernet protocol tells your computer to ignore everything that is not addressed to it, but that doesn't mean it has to. An advanced packet sniffer on an Ethernet network can look at all the traffic. Your neighbors could sniff everything on the Internet on your street. More advanced sniffers can trick the network in what is known as a "man-in-the-middle" attack. The sniffer appears to the router as the user's computer. All information is sent to the sniffer, which then copies the information before passing it on to the real addressee.

Many (but not most) websites now use a secure, encrypted connection when you log on so that your password is not sent in the clear for anyone sniffing around to pick up. Due to cost and speed, most then drop the connection back into an unsecure mode after the password transmission is made. When sniffing the transmission isn't possible, or when the data being sent is unreadable, that doesn't mean you are safe. A keystroke logger, a small hidden piece of malicious code installed surreptitiously on your computer, can capture everything you type and then transmit it secretly. Of course, this type of keystroke logger does require you to do something stupid in order for it to be installed on your computer, such as visiting a website that has been infected or downloading a file from an e-mail that is not really from someone you trust. In October 2008 the BBC reported that "computer scientists at the Security and Cryptography Laboratory at the Swiss Ecole Polytechnique Fédérale de Lausanne have demonstrated that criminals could use a radio antenna to 'fully or partially recover keystrokes' by spotting the electromagnetic radiation emitted when keys were pressed."

A fourth vulnerability of the Internet is its ability to propagate intentionally malicious traffic designed to attack computers, malware. Viruses, worms, and phishing scams are collectively known as "malware." They take advantage of both flaws in software and user errors like going to infected websites or opening attachments. Viruses are programs passed from user to user (over the Internet or via a portable format like a flash drive) that carry some form of payload to either disrupt a computer's normal operation, provide a hidden access point to the system, or copy and steal private information. Worms do not require a user to pass the program on to another user; they can copy themselves by taking advantage of known vulnerabilities and "worm" their way across the Internet. Phishing scams try to trick an Internet user into providing information such as bank account numbers and access codes by creating e-mail messages and phony websites that pretend to be related to legitimate businesses, such as your bank.

All this traffic is allowed to flow across the Internet with few, if any, checks on it. For the most part, you as an Internet user are responsible for providing your own protection. Most ISPs do not take even the most basic steps to keep bad traffic from getting to your computer, in part because it is expensive and can slow down the traffic, and also because of privacy concerns.

The fifth Internet vulnerability is the fact that it is one big network with a decentralized design. The designers of the Internet did not want it to be controlled by governments, either singly or collectively, and so they designed a system that placed a higher priority on decentralization than on security. The basic idea of the Internet began to form in the early 1960s, and the Internet as we know it today is deeply imbued with the sensibilities and campus politics of that era. While many regard the Internet as an invention of the military, it is really the product of now aging hippies on the campuses of MIT, Stanford, and Berkeley. They had funding through DARPA, the Defense Department's Advanced Research Project Agency, but the ARPANET, the Advanced Research Project Agency's Network, was not created just for the Defense Department to communicate. It initially connected four computers: at UCLA, Stanford, UC Santa Barbara, and, oddly, the University of Utah.

After building the ARPANET, the Internet's pioneers quickly moved on to figuring out how to connect the ARPANET to other networks under development. In order to do that, they developed the basic transmission protocol still used today. Robert Kahn, one of the ten or so people generally regarded as having created the Internet, laid out four principles for how these exchanges would take place. They are worth noting here now:

- Each distinct network should have to stand on its own, and no internal changes should be required to any such network to connect it to the Internet.
- Communications should be on a best-effort basis. If a packet didn't make it to the final destination, it should be retransmitted shortly from the source.
- Black boxes would be used to connect the networks; these would later be called gateways and routers. There should be no information retained by the gateways about the individual packets passing through them, thereby keeping them simple and avoiding complicated adaptation and recovery from various failure modes.
- There should be no global control at the operations level.

While the protocols that were developed based on these rules allowed for the massive growth in networking and the creation of the Internet as we know it today, they also sowed the seeds for the security problems. The writers of these ground rules did not imagine that anyone other than well-meaning academics and government scientists would use the Internet. It was for research purposes, for the exchange of ideas, not for commerce, where money would change hands, or for the purposes of controlling critical systems. Thus, it could be one network of networks, rather than separate networks for government, financial activity, etc. It was designed for thousands of researchers, not billions of users who did not know and trust each other.

Up to and through the 1990s, the Internet was almost universally seen as a force for good. Few of the Internet's boosters were willing to admit that the Internet was a neutral medium. It could easily be used to facilitate the free flow of communication between scientists and the creation of legitimate e-commerce, but could also allow terrorists to provide training tips to new recruits and to transmit the latest beheading out of Anbar Province on a web video. The Internet, much like the tribal areas of Pakistan or the tri-border region in South America, is not under the control of anyone and is therefore a place to which the lawless will gravitate.

Larry Roberts, who wrote the code for the first versions of the transmission protocol, realized that the protocols created an unsecure system, but he did not want to slow down the development of the new network and take the time to fix the software before deploying it. He had a simple answer for the concern. It was a small network. Rather than trying to write secure software to control the dissemination of information on the network, Roberts concluded that it would be far easier to secure the transmission lines by encrypting the links between each computer on the network. After all, the early routers were all in secure locations in government agencies and academic laboratories. If the information was secure as it traveled between two points on the network, that was all that really mattered. The problem was that the solution did not envision the expansion of the technology beyond the handful of sixty-odd computers that then made up the network. Trusted people ran all those sixty computers. A precondition for joining the network was that you were a known entity committed to promoting scientific advancement. And with so few people, if anything bad got on the network, it would not be hard to get it off and to identify who had put it there.

Then Vint Cerf left ARPA and joined MCI. Vint is a friend, a friend with whom I fundamentally disagree about how the Internet should be secured. But Vint is one of those handful of people who can legitimately be called "a father of the Internet," so what he thinks on Internet issues usually counts for a lot more than what I say. Besides, Vint, who always wears a bow tie, is a charming guy, and he now works for Google, which urges us all not to be evil.

MCI (now part of AT&T) was the first major telecommunications company to lay down a piece of the Internet backbone and to take the technology out of the small network of government scientists and academics, offering it to corporations and even, through ISPs, to home users. Vint took the transmission protocol with him, introducing the security problem to a far larger audience and to a network that could not be secured through encrypting the links. No one really knew who was connecting to the MCI network.

There are bound to be vulnerabilities in anything so large. Today, it has grown so extensive that the Internet is running out of addresses. When the Internet was cobbled together, the inventors came up with a numbering system to identify every device that would connect to the network. They decided that all addresses had to be a 32-bit number, a number so large that it would allow for 4.29 billion addresses. Never did they imagine that we would need more than that.

As of last count, there are nearly 6.8 billion people living on the planet. On the current standard, that's more than one address for every two people. And today, that is not enough. As the West grows more dependent on the Internet, and as the Second and Third worlds expand their use, 4.29 billion addresses cannot possibly satisfy all the possible people and devices that will want to connect to the web. That the Internet is running out of addresses on its own may be a manageable problem. If we move quickly to converting to the IPv6 address standard, by the time we run out of IPv4 addresses, in about two years, most devices should be able to operate on the new standard. But step back for a moment and a cause for concern begins to emerge.

The Pentagon envisions a near-future scenario in which every soldier on the battlefield will be a hub in a network, and as many as a dozen devices carried by that soldier will be plugged into the network and require their own addresses. If you stroll through the appliance aisle at a high-end home-goods store, you will notice that many of the washing machines, dryers, dishwashers, stoves, and refrigerators are advertising that they can be controlled through the Internet. If you are at work and want the oven to be preheated to 425 degrees when you arrive home, you could log onto a webpage, access your oven, and set it to the right temperature from your desktop.

What all this means is that as we move beyond 4.29 billion internal web addresses, the degree to which our society will be dependent on the Internet, for everything from controlling our thermostats to defending our nation, is set to explode, and with it the security problem is only going to get worse. What this could mean in a real-world conflict is something that until recently most policy makers in the Pentagon were loath to think about. It means that if you can hack into things on the Internet, you might not just be able to steal money. You might be able to cause some real damage, including damage to our military. So exactly how is it that you can hack into things, and why is that possible?

SOFTWARE AND HARDWARE

Of the three things about cyberspace that make cyber war possible, the most important may be the flaws in the software and hardware. All of those devices on the Internet we just discussed (the computer terminals and laptops, the routers and switches, the e-mail and web-page servers, the data files) are made by a large number of companies. Often, separate companies make the software that run devices. In the U.S. market, most laptops are made by Dell, HP, and Apple. (A Chinese company, Lenovo, is making a dent after having bought IBM's laptop computer unit.) Most big routers are made by Cisco and Juniper, and now the Chinese company Huawei. Servers are made by HP, Dell, IBM, and a large number of others, depending upon their purpose. The software they run is written mainly by Microsoft, Oracle, IBM, and Apple, but also by many other companies. Although these are all U.S. corporations, the machines (and sometimes the code that runs on them) come from many places.

In *The World Is Flat*, Thomas Friedman traces the production of his Dell Inspiron 600m notebook from the phone order he places with a customer-service representative in India to its delivery at his front door in suburban Maryland. His computer was assembled at a factory in Penang, Malaysia. It was "co-designed" by a team of Dell engineers in Austin and notebook designers in Taiwan. Most of the hard work, the design of the motherboard, was done by the Taiwanese team. For the rest of the thirty key components, Dell used a string of different suppliers. Its Intel processor might have been made in the Philippines, Costa Rica, Malaysia, or China. Its memory might have been made in Korea by Samsung, or by lesser-known companies in Germany or Japan. Its graphic card came from one of two factories in China. The motherboard, while designed in Taiwan, could have been made at a factory there, but probably came from one of two plants in Mainland China. The keyboard came from one of three factories in China, two of them owned by Taiwanese companies. The wireless card was made either by an American-owned company in China or by a Chinese-owned company in Malaysia or in Taiwan. The hard drive was probably made by the American company Seagate at a factory in Singapore, or by Hitachi or Fujitsu in Thailand, or by Toshiba in the Philippines.

After all these parts were assembled at the factory in Malaysia, a digital image of the Windows XP operating system (and probably Windows Office) was burned onto the hard drive. The code for that software, amounting to more than 40 million lines for XP alone, was written at a dozen or more locations worldwide. After the system was imprinted with the software, the computer was packaged up, placed on a pallet with 150 similar computers, and flown on a 747 to Nashville. From there, the laptop was picked up by UPS and shipped to Friedman. All told, Friedman proudly reports that "the total supply chain for my computer, including suppliers of suppliers, involved about four hundred companies in North America, Europe, and primarily Asia."

Why does Friedman spend six pages in a book about geopolitics documenting the supply chain for the computer he wrote the book on? Because he believes that the supply chain that built his computer knits together the countries that were part of that process in a way that makes interstate conflicts of the sort we saw in the twentieth century less likely. Friedman admits this is an update of his "Golden Arches Theory of Conflict Prevention" from his previous book, which argued that two states that both had a McDonald's would not go to war with each other. This time, Friedman's tongue-in-cheek argument has a little more meat to it than the hamburger theory. The supply chain is a microeconomic example of the trade that many theorists of international relations believe is so beneficial to the countries involved that even threatening war would not be worth the potential economic loss. Friedman looks at the averted crisis in 2004, when Taiwanese politicians running on a pro-independence platform were voted out of office. In his cute bumper-sticker-slogan way, Friedman observed that "Motherboards won over motherland," concluding that the status quo economic relationship was more valuable than independence to the Taiwanese voters.

Or maybe the Taiwanese voters just didn't want to end up dead after China invaded, which is what China more or less said it would do if Taiwan declared its independence. What Friedman sees as a force that makes conflict less likely, the supply chain for producing computers, may in fact make cyber warfare more likely, or at least make it more likely that the Chinese would win in any conflict. At any point in the supply chain that put together Friedman's computer (or your computer, or the Apple MacBook Pro that I am writing this book on), vulnerabilities were introduced, most accidentally, but probably some intentionally, that can make it both a target and a weapon in a cyber war.

Software is used as an intermediary between human and machine, to translate the human intention to find movie times online or read a blog, into something that a machine can understand. Computers really are just evolved electronic calculators. Early computer scientists realized that timed electrical pulses could be used to represent 1's and that the absence of a pulse could be used to represent 0's, like long and short bursts in Morse code. The base-10 numbers that humans use, because we have ten fingers, could be translated into this binary code that a machine could understand so that when, for instance, the 5 key on an early electronic calculator was depressed, it would close circuits that would send a pulse followed by a pause followed by another pulse in quick succession to represent the 1, 0, and 1 that make up the number 5 in a binary logic system.

All computers today are just evolutions of that same basic process. A simple e-mail message is converted into electric pulses that can be carried over copper wires and fiber-optic cables and then retranslated into a message readable to a human eye. To make that happen someone needed to provide instructions that a computer could understand. Those instructions are written in programming languages as computer code, and most people who write code make mistakes. The obvious ones get fixed, or else the computer program does not function as intended; but the less-obvious ones are often left in the code and can be exploited later to gain access. As computer systems have gotten faster, computer programs have grown more complex to take advantage of all the new speed and power. Windows 95 had less than 10 million lines of code. Windows XP, 40 million. Windows Vista, more than 50 million. In a little over a decade, the number of lines of code has grown by a factor of five, and with it the number of coding errors. Many of those coding errors allow hackers to make the software do something it was not supposed to, like let them in.

In order to manipulate popular software to do the wrong thing, like let you assume system administrator status, hackers design small applications, "applets," that are focused on specific software design or system configuration weaknesses and mistakes. Because computer crime is a big business, and getting ready to conduct cyber war is even well-funded, criminal hackers and cyber warriors are constantly generating new ways to trick systems. These hacker applications are called malware. On average in 2009, a new type or variant of malware was entering cyberspace every 2.2 seconds. Do the math. The three or four big antivirus software companies have sophisticated networks to look for the new malware, but they find and issue a "fix" for about one in every ten pieces of malware. The fix is a piece of software designed to block the malware. By the time the fix gets to the antivirus company's customers, often days, and sometimes weeks, have gone by. During that time, companies, government departments, and home users are entirely vulnerable to the new malware. They won't even know if they have been hit by it.

Frequently the malware is sitting on innocent websites, waiting for you. Let's say you surf to the website of a Washington think tank to read their latest analysis of some important public policy issue. Think tanks are notorious for not having enough money and not giving enough attention to creating secure and safe websites. So, as you are reading about the latest machinations over health care or human rights in China, a little piece of malware is downloading itself onto your computer. You have no way of knowing, but now your new friend in Belarus is logging your every keystroke. What happens when you log into your bank account or to the Virtual Private Network of your employer, the Really Big Defense Company? You can probably guess.

The most common software error for years, and one of the easiest to explain, is something called "buffer overflow." Code for a webpage is supposed to be written in such a way that when a user comes to that webpage, the user can only enter a certain amount of data, like a user name and password. It's supposed to be like Twitter, a program where you can enter, say, no more than 140 characters. But if the code writer forgets to put in the symbols that limit the number of characters, then a user can put in more. Instead of just putting in a user name or password, you could enter entire lines of instruction code. Maybe you enter instructions to allow you to add an account. Think about those instructions overflowing the limited area where a public user is supposed to be able to add information and then those instructions falling into the application. The instruction code reads as if a systems administrator had entered it and—ping!—you are inside.

Software errors are not easily discovered. Even experts cannot usually visually identify coding errors or intentional vulnerabilities in a few lines of code, let alone in millions. There is now software that checks software, but it is far from able to catch all the glitches in millions of lines. Each line of that code had to be written by a computer programmer, and each additional line of code increased the number of bugs introduced into the software. In some cases, programmers actually put those bugs in intentionally. The most famous case, and one that illustrates a larger phenomenon, occurred when somebody at Microsoft dumped an entire airplane-simulation program inside the Excel 97 database software. Microsoft only discovered it when people started thanking the company for it. Programmers may do it for fun, for profit, or in the service of a competing company or foreign intelligence service; but whatever their motive, it is a nearly impossible task to ensure that a few lines of code allowing for unauthorized access through a "trapdoor" are kept out of such massive programs. The original Trojan Horse had hidden commandos; today we have hidden commands of malicious code. In the case of the Excel spreadsheet, you began by opening a new blank document, pressing F5, and when a reference box opened, you typed in "X97:L97" and pressed *enter*, then pressed *tab*. This took you to cell M97 on the spreadsheet. Then if you clicked on the chart wizard button while holding down the *control* and *shift* keys—ping!—you activated a flight-simulator program, which popped right up.

Sometimes developers of code leave behind secret trapdoors so they can get back into the code easily later on to update it. Sometimes, unknown to their company, they do it for less reputable reasons. And sometimes other people, like hackers and cyber warriors, do it so they can get into parts of a network where they are not authorized. Thus, when someone hacks into a software product under development (or later), they may not just be stealing a copy, they may be adding to it. Intentional trapdoors, as well as others that occur because of mistakes in code writing, sometimes allow a hacker to gain what is called "root." Hackers trade or sell each other "root kits." If you have "root access" to a software program or a network, you have all the permissions and authorities of the software's creator or the network's administrator. You can add software. You can add user accounts. You can do anything. And, importantly, you can erase any evidence that you were ever there. Think of that as a burglar who wipes away his fingerprints and then drags a broom behind him to the door, erasing his footprints.

Code developers may go one step further than just leaving an access point and insert a "logic bomb." The term encompasses a spectrum of software applications, but the idea is simple. In addition to leaving behind a trapdoor in a network so you can get back in easily, without setting off alarms and without needing an account, cyber warriors often leave behind a logic bomb so they don't have to take the time to upload it later on when they need to use it. A logic bomb in its most basic form is simply an eraser, it erases all the software on a computer, leaving it a useless hunk of metal. More advanced logic bombs could first order hardware to do something to damage itself, like ordering an electric grid to produce a surge that fries circuits in transformers, or causing an aircraft's control surfaces to go into the dive position. Then it erases everything, including itself.

America's national security agencies are now getting worried about logic bombs, since they seem to have found them all over our electric grid. There is a certain irony here, in that the U.S. military invented this form of warfare. One of the first logic bombs, and possibly the first incidence of cyber war, occurred before there even really was much of an Internet. In the early 1980s, the Soviet leadership gave their intelligence agency, the KGB, a shopping list of Western technologies they wanted their spies to steal for them. A KGB agent who had access to the list decided he would rather spend the rest of his days sipping wine in a Paris café than freezing in Stalingrad, so he turned the list over to the French intelligence service in exchange for a new life in France. France, which was part of the Western alliance, gave it to the U.S. Unaware that Western intelligence had the list, the KGB kept working its way down, stealing technologies from a host of foreign companies. Once the French gave the list to the

CIA, President Reagan gave it the okay to help the Soviets with their technology needs, with a catch. The CIA started a massive program to ensure that the Soviets were able to steal the technologies they needed, but the CIA introduced a series of minor errors into the designs for things like stealth fighters and space weapons.

Weapons designs, however, were not at the top of the KGB's wish list. What Russia really needed was commercial and industrial technology, particularly for its oil and gas industry. In order to get the product from the massive reserves in Siberia to Russian and Western consumers, oil and gas had to be piped over thousands of miles. Russia lacked the technology for the automated pump and valve controls crucial to managing a pipeline thousands of miles long. They tried to buy it from U.S. companies, were refused, and so set their sights on stealing it from a Canadian firm. With the complicity of our northern neighbors, the CIA inserted malicious code into the software of the Canadian firm. When the Russians stole the code and used it to operate their pipeline, it worked just fine, at least initially. After a while, the new control software started to malfunction. In one segment of the pipeline, the software caused the pump on one end to pump at its maximum rate and the valve at the other end to close. The pressure buildup resulted in the most massive non-nuclear explosion ever recorded, over three kilotons.

If the Cold War with Russia heats up again, or if we were to go to war with China, this time it might be our adversaries who have the upper hand in cyber war. The United States' sophisticated arsenal of space-age weapons could be turned against us to devastating effect. Our air, land, and sea forces rely on networked technologies that are vulnerable to cyber weapons that China and other near peer adversaries have developed with the intention of eliminating our conventional superiority. The U.S. military is no more capable of operating without the Internet than Amazon.com would be. Logistics, command and control, fleet positioning, everything down to targeting, all rely on software and other Internet-related technologies. And all of it is just as insecure as your home computer, because it is all based on the same flawed underlying technologies and uses the same insecure software and hardware.

With the growth of outsourcing to countries like India and China that Friedman got so excited about, the likelihood that our peer competitors have been able to penetrate major software and hardware companies and insert such code into the software we rely on has only increased. In the world of computer science and networking, experts long thought that the two most ubiquitous operating-system codes (software that tells hardware what to do) were also the most badly written, or "buggy," computer code. They were Microsoft's Windows operating system for desktop and laptop computers, and Cisco's for large Internet routers. Both systems were proprietary, meaning not publicly available. You could buy the software as a finished product, but you could not get the underlying code. There were, however, several known instances in which Microsoft's security was compromised and the code stolen, giving the recipient the opportunity to identify the software errors and ways to exploit them.

I mentioned above (in chapter 2) that China had essentially blackmailed Microsoft into cooperating with it. China had announced that it would develop its own system based on Linux, called Red Flag, and said it would require that it be used instead of Microsoft. Soon Microsoft was bargaining with the Chinese government at the highest level, helped along by its consultant, Henry Kissinger. Microsoft dropped its price, gave the Chinese its secret code, and established a software research lab in Beijing (the lab is directly wired into Microsoft's U.S. headquarters). A deal was struck. It must have been a good deal: the President of China then visited Bill Gates at his home near Seattle. The Chinese government now uses Microsoft, but it is that special variation with a Chinese government encryption module. One former U.S. intelligence officer told us, "This may mean that no one can hack Windows easily to spy on China. It certainly does not mean that China is less able to hack Windows to spy on others."

What can be done to millions of lines of code can also be done with millions of circuits imprinted on computer chips inside computers, routers, and servers. Chips are the guts of a computer, like software in silicon. They can be customized, just like software. Most experts cannot look at a complicated computer chip and determine whether there is an extra piece here or there, a physical trapdoor. Computer chips were originally made in the U.S., although now they are mostly manufactured in Asia. The U.S. government once had its own chip factory, called a "fab" (short for "fabrication facility"); however, the facility has not kept pace with technology and cannot manufacture the chips required for modern systems. Recently the world's second-largest chip manufacturer, AMD, announced its intentions to build the most advanced fab in the world in upstate New York. It will be partially government funded, but not by the U.S. government: AMD got a big investment from the United Arab Emirates.

It is not that the U.S. government is unaware of the problem of software and hardware being made globally. In fact, in his last year in office, President George W. Bush signed PDD-54, a secret document that outlines steps to be taken to defend the government better from cyber war. One of those programs is reported to be a "Supply Chain Security" initiative, but it will be difficult for the U.S. government to purchase only software and hardware made in the U.S. under secure conditions. Currently, it would be difficult to find any.

MACHINES CONTROLLED FROM CYBERSPACE

Neither the vulnerabilities of Internet design nor the flaws in software and hardware quite explain how cyber warriors could make computers attack. How is it that some destructive hand can reach out from cyberspace into the real world and cause serious damage?

The answer stems from the rapid adoption of the Internet and cyberspace by industries in the U.S. in the 1990s. During that decade evangelical information-technology companies showed other corporations how they could save vast amounts of money by taking advantage of computer systems that could do things deep into their operations. Far beyond e-mail or word processing, these business practices involved automated controls, inventory monitoring, just-in-time delivery, database analytics, and limited applications of artificial-intelligence programs. One Silicon Valley CEO told me enthusiastically in the late 1990s how he had applied these techniques to his own firm. "Somebody wants to buy something, they go online to our site. They customize the product they want and hit BUY. Our system notifies the parts makers, plans to ship the parts to the assembly plant, and schedules assembly and delivery. At the assembly plant, robotic devices put the product together and put it in a box with a delivery label on it. We don't own the computer server that took the order, the parts plants, the assembly plant, or the delivery aircraft and trucks. It's all outsourced and it's all just-in-time delivery." What he owned was the research department, the design team, and some corporate overhead. At companies like his, and in the U.S. economy in general, profitability soared.

What made all of that possible was the deep penetration in the 1990s of information-technology systems into companies, into every department. In many industries, controls that were once manually activated were converted to digital processors. Picture the factory or plant of the twentieth century where some guy in a hard hat got a call from his supervisor telling him to go over and crank some round valve or change some setting. I can see it vividly, my father worked in a place like that. Today, in almost every industry, fewer people are required. Digital control systems monitor activity and send commands to engines, valves, switches, robotic arms, lights, cameras, doors, elevators, trains, and aircraft. Intelligent inventory systems monitor sales in real time and send out the orders to make and ship replacements, often without a human in the loop.

The conversion to digital control systems and computer-managed operations was quick and thorough. By the turn of the century, most of the old systems were retired, even from the role of "backup." Like Cortés burning his ships after arriving in the New World, U.S. companies and government agencies built a new world in which there were only computer-based systems. When the computers fail, employees stand around doing nothing or go home. Try to find a typewriter and you will get the picture of this new reality.

Just as the Internet, and cyberspace in general, is replete with software and hardware problems and configuration shortcomings, so are the computer networks that run major corporations, from utilities to transportation to manufacturing. Computer networks are essential for companies or government agencies to operate. "Essential" is a word chosen with care, because it conveys the fact that we are dependent upon computer systems. Without them, nothing works. If they get erroneous data, systems may work, but they will do the wrong things.

Despite all the money spent on computer security systems, it is still very possible to insert erroneous data into networks. It can mean that systems shut down, or damage themselves, or damage something else, or send things or people to the wrong places. At 3:28 p.m. on June 11, 1999, a pipeline burst in Bellingham, Washington. Gasoline began spilling out into the creek below. The gas quickly extended well over a mile along the creek. Then it caught fire. Two ten-year-old boys playing along the stream were killed, as was an eighteen-year-old farther up the creek. The nearby municipal water-treatment plant was severely damaged by the fire. When the U.S. National Transportation Safety Board examined why the pipeline burst, it focused on "the performance and security of the supervisory control and data acquisition (SCADA) system." In other words, the software failed. The report does not conclude that in this case the explosion was intentionally caused by a hacker, but it is obvious from the analysis that pipelines like the one in Bellingham can be manipulated destructively from cyberspace.

The clearest example of the dependency and the vulnerability brought on by computer controls also happens to be the one system that everything else depends upon: the electric power grid.

As a result of deregulation in the 1990s, electric power companies were divided up into generating firms and transmission companies. They were also allowed to buy and sell power to each other anywhere within one of the three big power grids in North America. At the same time, they were, like every other company, inserting computer controls deep into their operations. Computer controls were also installed to manage the buying and selling, generation, and transmission. A SCADA system was already running each electric company's substations, transformers, and generators. That Supervisory Control and Data Acquisition system got and sent signals out to all of the thousands of devices on the company's grid. SCADAs are software programs, and most electric companies use one of a half dozen commercially available products.

These control programs send signals to devices to regulate the electric load in various locations. The signals are most often sent via internal computer network and sometimes by radio. Unfortunately, many of the devices also have other connections, multiple connections. One survey found that a fifth of the devices on the electric grid had wireless or radio access, 40 percent had connections to the company's internal computer network, and almost half had direct connections to the Internet. Many of the Internet connections were put in place to permit their manufacturers to do remote diagnostics.

Another survey found that at one very large electric company, 80 percent of the devices were connected to the corporate intranet, and there were, of course, connections from the intranet out to the public Internet. What that means is that if you can hack from the Internet to the intranet, you can give orders to devices on the electric grid, perhaps from some nice cyber café on the other side of the planet. Numerous audits of electric power companies by well-respected cyber security experts have found that this is all very doable. What sort of things might you do with controls to the grid?

In 2003, the so-called Slammer worm (big, successful computer malware attacks get their own names) got into and slowed controls on the power grid. A software glitch in a widely used SCADA system also contributed to the slowed controls. So when a falling tree created a surge in a line in Ohio, the devices that should have stopped a cascading effect did not do so until the blackout got to somewhere in southern New Jersey. The result was that eight states, two Canadian provinces, and 50 million people were without electricity, and without everything that needs electricity (such as the water system in Cleveland). The tree was the initiator, but the same effects could have been achieved by a command given over the control system by a hacker. In fact, in 2007 CIA expert Tom Donahue was authorized to tell a public audience of experts that the Agency was aware of instances when hackers had done exactly that. Although Tom didn't say where hackers had caused a blackout as part of a criminal scheme, it was later revealed that the incident took place in Brazil.

The 2003 blackout lasted a few long hours for most people, but even without anyone trying to prolong the effect it lasted four days in some places. In Auckland, New Zealand, in 1998 the damage from overloading power lines triggered a blackout and kept the city in the dark for five weeks. If a control system sends too much power down a high-tension line, the line itself can be destroyed and initiate a fire. In the process, however, the surge of power can overwhelm home and office surge protectors and fry electronic devices, computers to televisions to refrigerators, as happened recently in my rural county during a lightning storm.

The best example, however, of how computer commands can cause things to destroy themselves may be electric generators. Generators make electricity by spinning, and the number of times they spin per minute creates power in units expressed in a measurement called Hertz. In the United States and Canada, the generators on most subgrids spin at 60 Megahertz. When a generator is started, it is kept off the grid until it gets up to 60 MHz. If it is connected to the grid at another speed, or if its speed changes very much while on the grid, the power from all of the other generators on the grid spinning at 60 MHz will flow into the slower generator, possibly ripping off its turbine blades.

To test whether a cyber warrior could destroy a generator, a federal government lab in Idaho set up a standard control network and hooked it up to a generator. In the experiment, code-named Aurora, the test's hackers made it into the control network from the Internet and found the program that sends rotation speeds to the generator. Another keystroke and the generator could have severely damaged itself. Like so much else, the enormous generators that power the United States are manufactured when they are ordered, on the just-in-time delivery principle. They are not sitting around, waiting to be sold. If a big generator is badly damaged or destroyed, it is unlikely to be replaced for months.

Fortunately, the Federal Electric Regulatory Agency in 2008 finally required electric companies to adopt some specific cyber security measures and warned that it would fine companies for noncompliance up to one million dollars a day. No one has been fined yet. The companies have until sometime in 2010 to comply. Then the commission promises it will begin to inspect some facilities to determine if they are compliant. Unfortunately, President Obama's "Smart Grid" initiative will cause the electric grid to become even more wired, even more dependent upon computer network technology.

The same way that a hand can reach out from cyberspace and destroy an electric transmission line or generator, computer commands can derail a train or send freight cars to the wrong place, or cause a gas pipeline to burst. Computer commands to a weapon system may cause it to malfunction or shut off. What a cyber warrior can do, then, is to reach out from cyberspace, causing things to shut down or blow up, things like the power grid, or a thousand other critical systems, things like an opponent's weapons.

The design of the Internet, flaws in software and hardware, and allowing critical machines to be controlled from cyberspace, together, these three things make cyber war possible. But why haven't we fixed these problems by now?

CHAPTER FOUR

THE DEFENSE FAILS

Thus far we have seen evidence that there have been "trial runs" at cyber war, mostly using primitive denial of service attacks. We have seen how the United States, China, Russia, and others are investing heavily in cyber war units. We have imagined what the first few minutes of a devastating, full-scale cyber attack on the U.S. would look like. And we have walked through what it is about cyber technology and its uses that makes such a devastating attack possible.

Why hasn't anybody done anything to fix these vulnerabilities? Why are we placing such emphasis on our ability to attack others, rather than giving priority to defending ourselves against such an attack? People have tried to create a cyber war defense for the U.S. Obviously they have not succeeded. In this chapter we'll review what efforts have been made to defend against cyber war (and cyber crime, and cyber espionage) and see why they have been such an unmitigated failure. Strap yourself in, we are first going to move quickly through twenty years of efforts in the U.S. to do something about cyber security. Then we will talk about why it hasn't worked.

INITIAL THOUGHTS AT THE PENTAGON

In the early 1990s the Pentagon began to worry about the vulnerability created by reliance on new information systems to conduct warfare. In 1994, something called the "Joint Security Commission" that was set up by DoD and the intelligence community focused on the new problem introduced by the spread of networked technology. The commission's final report got three important concepts right:

- "Information systems technology . . . is evolving at a faster rate than information systems security technology."
- "The security of information systems and networks [is] the major security challenge of this decade and possibly the next century and . . . there is insufficient awareness of the grave risks we face in this arena."
- The report also noted that the increased dependence in the private sector on information systems made the nation as a whole, not just the Pentagon, more vulnerable.

These three points are all true and even more relevant today. A prescient *Time* magazine article from 1995 demonstrates the point that cyber war and domestic vulnerabilities were subjects to which Washington was alerted fifteen years ago. We keep rediscovering this wheel. In the 1995 story, Colonel Mike Tanksley waxed poetic about how in a future conflict with a lesser power the United States would force our enemy to submit without our ever having fired a shot. Using hacker techniques that were then only possible in the movies, Colonel Tanksley described how America's cyber warriors would take down the enemy's phone system, destroy the routing system for the country's rail line, issue phony commands to the opposing military, and take over television and radio broadcasts to flood them with propaganda. In the fantasy scenario that Tanksley describes, the effect of using these tactics would end the conflict before it even starts. *Time* magazine reported that a logic bomb "would remain dormant in an enemy system until a predetermined time, when it would come to life and begin eating data. Such bombs could attack, for example, computers that run a nation's air defense system or central bank." The article told readers that the CIA had a "clandestine program that would insert booby-trapped computer chips into weapons systems that a foreign arms manufacturer might ship to a potentially hostile country—a technique called 'chipping.'" A CIA source told the reporters how it was done, explaining, "You get into the arms manufacturer's supply network, take the stuff offline briefly, insert the bug, then let it go to the country. . . . When the weapons system goes into a hostile situation, everything about it seems to work, but the warhead doesn't explode."

The *Time* article was a remarkable piece of journalism that captured both complicated technical issues and the resulting policy problems long before most people in government understood anything about them. On the cover it asked: "The U.S. rushes to turn computers into tomorrow's weapons of destruction. But how vulnerable is the homefront?" That question is as pertinent today as it was then, and, remarkably, the situation has changed very little. "An infowar arms race could be one the US would lose because it is already so vulnerable to such attacks," the writers conclude. "Indeed," they continue, "the cyber enhancements that the military is banking on for its conventional forces may be chinks in America's armor." So by the mid-1990s journalists were seeing that the Pentagon and the intelligence agencies were excited about the possibility of creating cyber war capabilities, but doing so would create a double-edged sword, one that could be used against us.

MARCHING INTO THE MARSH

Timothy McVeigh and Terry Nichols woke a lot of people up in 1995. Their inhumane attack in Oklahoma City, killing children at a day care center and civil servants at their desks, really got to Bill Clinton. He delivered an especially moving eulogy near the site of the attack. When he came back to the White House, I met with him, along with other White House staff. He was thinking concep-tually, as he often does. Society was changing. A few people could have significant destructive power. People were blowing things up in the U.S., not just in the Middle East. What if the truck bomb had been aimed at the stock market, or the Capitol, or some building whose importance we didn't even recognize? We were becoming a more technological nation, but in some ways that also was making us a more fragile nation. At the urging of Attorney General Janet Reno, Clinton appointed a commission to look at our vulnerability as a nation to attacks on important facilities.

Important facilities got translated into bureaucratese as "critical infrastructure," a phrase that continues, and continues to confuse, today. The new panel got the moniker Presidential Commission on Critical Infrastructure Protection (PCCIP). Not surprisingly, then, most people referred to it using the name of its Chairman, retired Air Force General Robert Marsh. The Marsh Commission was a full-time endeavor for a large panel and a professional staff. They held meetings throughout the country and talked to experts in numerous industries, universities, and government agencies. What they came back with in 1997 was not what we expected. Rather than focusing on right-wingers like McVeigh and Nichols or al Qaeda terrorists like those who had attacked the World Trade Center in 1993, Marsh sounded a loud alarm about the Internet. Noting what was then a recent trend, the Marsh Commission said that important functions from rail to banking, from electricity to manufacturing were all being connected to the Internet and yet that network of networks was completely insecure. By hacking from the Internet, an attacker could shut down or damage "critical infrastructure."

Raising the prospect of nation-states creating "information war" attack units, Marsh called for a massive effort to protect the U.S. He identified the chief challenge as being the role of the private sector, which owned most of what counted as "critical infrastructure." Industries were wary of the government regulating them to promote cyber security. Instead of doing that, Marsh called for a "public-private partnership," heightened awareness, sharing of information, and research into more secure designs.

I was disappointed, although in time I came to understand that General Marsh was right. As the senior White House official in charge of security and counterterrorism issues, I had hoped for a report that would have helped me get the funding and structure I needed to deal with al Qaeda and others. Instead, Marsh was talking about computers, which was not my job. My close friend Randy Beers, then Special Assistant to the President for Intelligence and the man who had been shepherding the Marsh Commission for the White House, walked next door to my office (with its twenty-foot-high ceiling and great view of the National Mall), plunked himself down in a chair, and announced, "You have to take over critical infrastructure. I can't do it because of the Clipper chip."

The Clipper chip had been a plan, developed in 1993 by NSA, in which the government would require anyone in the U.S. using encryption to install a chip that would let NSA listen in, with a court order. Privacy, civil liberties, and technology interest groups united in vehement opposition. For some reason, they did not trust that NSA would only listen in when they had a warrant (which, under George W. Bush, later proved to be true). The Clipper chip got killed by 1996, but it had left a lot of distrust between the growing information technology (IT) industry and the U.S. intelligence community. Beers, being an intelligence guy, felt he could not gain the trust of the IT industry. So he dumped it in my lap. Moreover, he had already wired that decision with the National Security Advisor, Sandy Berger, who asked me to write a Presidential Decision Document stating our policy on the issue, and putting me in charge of it.

The result was a clear statement of the problem and our goal, but within a structure with limitations that prevented us from achieving it. The problem was that "because of our military strength, future enemies . . . may seek to harm us . . . with non-traditional attacks on our infrastructure and information systems . . . capable of significantly harming both our military power and our economy." So far so good. The goal was that "any interruptions or manipulations of critical functions must be brief, infrequent, manageable, geographically isolated, and minimally detrimental." Pretty good stuff.

But how to do it? By the time every agency in government had watered the decision down, it read: "The incentives that the market provides are the first choice for addressing the problem of critical infrastructure protection. . . . [We will consider] regulation only in the event of a material failure of the market . . . [and even then] agencies will identify alternatives to direct regulation." I got a new title in the Decision Document, but it would not fit on a business card: "National Coordinator for Security, Infrastructure Protection, and Counter-terrorism." Little wonder the media used the term "czar"; no one could remember the real title. The Decision Document made clear, however, that the czar could not direct anyone to do anything. The Cabinet members had been adamant about that. No regulation and no decision-making authority meant little potential for results.

Nonetheless, we set off to work with the private sector and with government agencies. The more I worked on the issue, the more concerned I became. Marsh had not really been alarmist, I came to appreciate; he and his commission had actually understated the problem. Our work on the Y2K computer glitch (the fact that most software could not roll over from 1999 to 2000 and might, therefore simply freeze up) greatly added to my understanding of just how much everything was rapidly becoming dependent upon computer-controlled systems and networks connected in some way to the Internet. In the 2000 federal budget, I was able to add $2 billion for improved cyber security efforts, but it was a fraction of what was needed.

By 2000, we had developed a National Plan for Information Systems Protection, but there was still no willingness in the government to attempt to regulate the industries that ran the vulnerable critical infrastructure. To highlight the ideological correctness of the decision to avoid regulation, I used the phrase "eschew regulation" in the decision document, mimicking Maoist rhetoric. (Mao had directed, "dig tunnels deeper, bury food everywhere, eschew hegemonism.") No one saw the irony. Nor would the Cabinet departments even do enough to protect their own networks, as called for in the Presidential Directive. Thus, the plan was toothless. It did, however, make clear to industry and to the public what the stakes were. Bill Clinton's cover letter left no doubt that the IT revolution had changed how the economy and national defense were done. From turning on the lights, to calling 911, to boarding an aircraft, we now relied upon computer-driven systems. A "concerted attack" on the computers of an important economic sector would "have catastrophic results." This was not a theoretical potential; rather, "we know the threat is real." Opponents that had relied on "bombs and bullets" could now use "a laptop . . . [as] a weapon capable of . . . enormous damage."

I added in a cover letter of my own that "More than any other nation, America is dependent upon cyberspace." Cyber attacks could "crash electric grids . . . transportation systems . . . financial institutions. We know other governments are developing that capability." So were we, but I didn't say that.

SIX FUNNY NAMES

During those initial years of my focusing on cyber security there were six major incidents that convinced me that this was a serious problem. First, in 1997, I worked with NSA on a test of the Pentagon's cyber security in an exercise the military called "Eligible Receiver." Within two days, our attack team had penetrated the classified command network and was in position to issue bogus orders. I stopped the exercise early. The Deputy Defense Secretary was shocked at the Pentagon's vulnerability and ordered all components to buy and install intrusion detection systems. They quickly discovered that there were thousands of attempts a day to hack into DoD networks. And those were the ones they knew about.

In 1998, during a crisis with Iraq, someone hacked into the unclassified DoD computers that were needed to manage the U.S. military buildup. The FBI gave the attack the appropriate name "Solar Sunrise" (it was a wake-up call for many). After a few days of panic, the attackers were discovered to be not Iraqi but Israeli. Specifically, a teenager in Israel and two more in California had proved how poorly secured our military logistics network was.

In 1999, an Air Force base noticed something odd about its computer network. The Air Force called the FBI, which called NSA. What emerged was that huge amounts of data were being exfiltrated from the research files at the airbase. Indeed, gigantic amounts of data were being shipped out from a lot of computers in the Defense network and from many data systems in the national nuclear laboratories of the Energy Department. The FBI case file for this one was called "Moonlight Maze," which also turned out to be apt because no one could throw much light on what was happening other than to say the data was being sent through a long series of stops in many countries before ending up somewhere. The two deeply disturbing aspects of this were that the computer security specialists could not stop the data from being stolen, even when they knew about the problem, and no one was really sure where it all was going (although some people later publicly attributed the attack to Russians). Every time new defenses were put in place, the attacker beat them. Then, one day, the attacks stopped. Or, more likely, they started attacking in a way we could not see.

Early in 2000, when we were still glowing from our success in avoiding a Y2K problem, a number of the new Internet commerce sites (AOL, Yahoo, Amazon, E-Trade) crashed from what I was told was a DDOS, a term new to most people in 2000. This was the first "big one," hitting numerous companies simultaneously and knocking them down. The motive was hard to discern. There were no monetary demands, nor was there a real political message. Somebody seemed to be trying out the concept of covertly taking over lots of people's computers and secretly using them to attack. (That somebody later turned out to be a busboy from Montreal.) I saw the DDOS as an opportunity to have the government remind the private sector that they needed to take cyber attacks seriously.

President Clinton agreed to host the leaders of the companies that had been attacked as well as other CEOs from important infrastructures and from the IT industry. It was the first presidential White House meeting with private-sector leadership concerning a cyber attack. It was also the last, thus far. Although it was a remarkably detailed and frank meeting, eye-opening for many, it essentially resulted in everyone agreeing to work harder on the problem.

In 2001, the new Bush Administration got a taste of the problem when the Code Red worm infected over 300,000 computers in a few hours and then turned them all into zombies programmed to launch a DDOS attack on the White House webpage. I was able to distribute the White House website onto 20,000 servers using a company called Akamai and thereby avoided the effects of the attack (we also persuaded some of the major ISPs to divert the attack traffic). Cleaning up the infected computers turned out to be a harder job. Many companies and individuals could not be bothered to remove the worm software, despite its repeated disruptive effects on the Internet. Nor did we have any ability to deny those machines access to the Internet, even though they were pumping out malware on a regular basis. In the days after the 9/11 terrorist attack, another, more serious worm spread quickly. The NIMDA (Admin spelled backward) worm was targeted at computers running in the most well secured private-sector industry vertical, the financial industry. Despite their sophisticated security, many banks and Wall Street firms were knocked offline.

CYBER SECURITY GETS BUSHED

The Bush Administration took some convincing that cyber security was an important problem, but agreed by the summer of 2001 to set up a separate office in the White House to handle its coordination (Executive Order 13231). I ran that office as Special Advisor to the President for Cybersecurity from the autumn of 2001 to early 2003. Most of the rest of the Bush White House (the Science Advisor, the Economic Advisor, the Budget Director) sought to limit the authority of the new cyber security position.

Unfazed by that, my team took the Clinton National Plan and modified it based on input from twelve industry teams we established and from citizen input at ten town halls held around the country. (The kind of crowd that shows up for a cyber security town hall is, thankfully, more civilized than the nut jobs who showed up in 2009 at health-care town halls.) The result was the National Strategy to Secure Cyberspace, which Bush signed in February 2003. Substantively, there was little difference between the Clinton and Bush approaches, except that the Republican administration not only continued to eschew regulation, they downright hated the idea of the federal government issuing any new regulations on anything at all. Bush left jobs vacant for long periods at several regulatory commissions and then appointed commissioners who did not enforce the regulations that did exist.

Bush's personal understanding and interest in cyber security early in his administration were best summed up by a question he asked me in 2002. I had gone to him in the Oval Office with news of a discovery of a pervasive flaw in software, a flaw that would allow hackers to run amok unless we could quietly persuade most major networks and corporations to fix the flaw. Bush's only reaction was: "What does John think?" John was the CEO of a large information-technology company and a major donor to the Bush election committee.

With the creation of the Department of Homeland Security, I had thought there would be an opportunity to take many of the scattered entities working on cyber security and merge them into one center of excellence. As a result, some cyber security offices from the Commerce Department, FBI, and DoD were brought together in Homeland. The sum turned out to be much less than the parts, as many of the best people in the merged offices took the opportunity to leave government. When I also took my exit from the Bush Administration shortly before it began the disastrous Iraq War, the

White House chose not to replace me as Special Advisor. The most senior official in government charged with coordinating cyber security was then in an office buried several layers down in what was turning into the most dysfunctional department in government, DHS. Several very good people tried to make that job work, but each one quit in frustration. The media began talking about the "cyber czar of the week." The high-level private-sector focus on the issue we had achieved faded.

Four years later, Bush made a decision much more quickly than his staff had assumed he would. There was a covert action that the President had to approve personally. The President's scheduler had booked an hour for the decision briefing. It took five minutes. Bush never saw a covert-action proposal he didn't like. Now, with fifty-five minutes left in the meeting, the Director of National Intelligence, Mike McConnell, saw an opening. All the right people were in the room, senior national security cabinet members. McConnell asked if he could discuss a threat to the financial industry and the U.S. economy. Given the floor, he talked about cyber war and how vulnerable we were to it. Particularly vulnerable was the financial sector, which would not know how to recover from a data-shredding attack, an attack that could do unimaginable damage to the economy. Stunned, Bush turned to Treasury Secretary Hank Paulson, who agreed with the assessment.

At this point, Bush, who had been sitting behind the large desk in the Oval Office, almost jumped in the air. He moved quickly to the front of the desk and began gesturing for emphasis as he spoke. "Information technology is supposed to be our advantage, not our weakness. I want this fixed. I want a plan, soon, real soon." The result was the Comprehensive National Cybersecurity Initiative (CNCI) and National Security Presidential Decision 54. Neither has ever become public. Both documents call, appropriately enough, for a twelve-step plan. They focus, however, on securing the government's networks. Oddly, the plan did not address the problem that had started the discussion in the Oval Office, the vulnerability of the financial sector to cyber war.

Nonetheless, Bush requested $50 billion over five years for the Comprehensive National Cybersecurity Initiative, which is neither comprehensive nor national. The initiative is an effort to, in the words of one knowledgeable insider, "stop the bleeding" out of DoD and intelligence-community systems, with a secondary focus on the rest of the government. Also described as a multibillion-dollar "patch and pray program," the initiative does not address vulnerabilities in the private sector, including in our critical infrastructures. That tougher problem was left to the next administration.

The initiative was also supposed to develop an "information warfare deterrence strategy and declaratory doctrine." That part has almost totally been put on hold. In May 2008, the Senate Armed Services Committee criticized the initiative's secrecy in a public report, with the comment that "it is difficult to conceive how the United States could promulgate a meaningful deterrence doctrine if every aspect of our capabilities and operational concepts is classified." Reading that, I could not help but think of Dr. Strangelove when, in the movie of the same name, he berates the Soviet Ambassador for Moscow's keeping the existence of its nuclear-deterrent Doomsday Machine a secret: "Of course, the whole point of a Doomsday Machine is lost if you *keep it a secret*! Why didn't you tell the world?" The reason we are keeping our cyber deterrence strategy secret is probably that we do not have a good one.

OBAMA'S OVERFLOWING PLATE

It was another vulnerability of the financial sector, brought on as a result of industry successfully lobbying against government regulation, that Barack Obama was forced to focus on when he became President in 2009. The subprime-mortgage meltdown and the complex dealings in the derivatives markets had created the worst financial crisis since 1929. With that, in addition to the war in Iraq, the war in Afghanistan, threatening flu pandemics, health-care reform, and global warming all requiring his attention, Obama did not focus on cyber security. He had, however, addressed the issue during the 2008 campaign. Although I had signed on to the campaign as a terrorism advisor, I used that access to pester the candidate and his advisors about cyber war. It was not surprising to me that Obama "got" the issue, since he was running the most technologically advanced, cyber-dependent presidential campaign in history.

Thus, as part of the campaign's effort to stake out some ground on national security issues, then-Senator Obama gave a speech and met with national experts on technology and emerging threats at Purdue University in the summer of 2008. In the speech, he took the bold step of declaring U.S. cyber infrastructure "a strategic asset," an important phrase in government-speak that means it is something worth defending. He also pledged to appoint a senior White House advisor who would report directly to him and gave a general commitment to make cyber security "a top federal priority." In the accompanying fact sheet, which my coauthor Rob Knake drafted along with two MIT computer scientists, John Mallery and Roger Hurwitz, he went a step further, criticizing the Bush Administration for moving too slowly in the face of the risks associated with cyberspace, and pledging to initiate a "Safe Computing R&D effort" to "develop next-generation secure computers and networking for national security applications," to invest more in science and math education, and to create plans to address private-sector vulnerabilities, identity theft, and corporate espionage.

A few weeks later, the cyber threat was hammered home to Obama in a very serious way. The FBI quietly informed the campaign that it had reason to believe Chinese hackers had infiltrated the campaign's computer systems. I asked one of my business partners, Paul Kurtz (who had worked on cyber security on both the Clinton and Bush White House staffs), to take a team of cyber security experts out to the Chicago headquarters to assess the extent of the damage and see what could be done to secure the systems. The Chinese hackers had focused on draft policy documents. They had used some sophisticated techniques, hidden beneath more obvious activity.

When the campaign quietly put together an unofficial transition team weeks before election day, I asked everyone working on national security planning to stop using their home computers for that purpose. Even though what they were writing was unclassified, it was of interest to China and others (including, presumably, John McCain, not that his campaign had shown much understanding of cyber technology). With the campaign's blessing, we distributed "clean" Apple laptops and locked them down so they could only connect to one thing, a virtual private network we created using a server with a completely innocuous name. I knew we were going to be in trouble when I started getting calls complaining about the security features. "Dick, I'm at a Starbucks and this damn machine won't let me connect to the wi-fi." "Dick, I want to pull some files off of my Gmail account, but I can't access the Internet." I tried to point out that if you are a senior member of the informal national security transition team, you probably should not be planning the takeover of the White House from a Starbucks, but not everyone seemed to care.

Shortly before the inauguration, Paul Kurtz and I provided the new White House team with a draft decision document to formalize the proposals Obama had advocated in the Purdue speech. We argued that if Obama waited, people would come out of the woodwork to try to stop it. Although the most senior White House staff understood that problem and wanted a quick decision, it was, understandably, not a high priority for them. Instead, the new Obama White House announced a Sixty Day Review and asked one of the drafters of Bush's CNCI to run it. This was despite the fact that Jim Lewis and the Commission on Cyber Security for the forty-fourth Presidency had already spent over a year working to achieve a consensus view on what the next President needed to do, releasing their report on December 8, 2008. When, 110 days later, the President announced the results, guess what? It was CNCI redux. It also had a military Cyber Command, but not a cyber war strategy, not a major policy or program to defend the private sector, nothing to initiate international dialogue on cyber war. And, déjà vu all over again, the new Democratic President went out of his way to take regulation off the table: "So let me be very clear: my administration will not dictate security standards for private companies."

What Obama did not announce in his public remarks after the Sixty Day Review was who would be the new White House cyber security czar. Few qualified people wanted the job, largely because it had no apparent authority and had been altered to report directly to both the Economic Advisor and the National Security Advisor. The Economic Advisor was the ousted former Harvard president Larry Summers, who had made it clear that he thought the private sector and market forces would do enough to deal with the cyber war threat without any additional government regulation or role in their affairs. Months went by during which the best efforts of the White House personnel office failed to convince candidate after candidate that this was a job worth taking.

Thus, for the first year of his administration, Obama had no one in the White House trying to orchestrate a government-wide, integrated cyber security or cyber war program. Departments and agencies did their own thing, or did nothing. The two lead agencies in defending America from cyber war were U.S. Cyber Command (to defend the military) and the Department of Homeland Security (to defend, well, something else). The head of U.S. Cyber Command kept a low profile for most of 2009 because the Senate had not yet agreed to give him his fourth star. To get the promotion from three stars, General Keith Alexander would have to answer questions before a Senate committee, and that committee wasn't too sure it understood what U.S. Cyber Command was actually supposed to do. Senator Carl Levin of Michigan asked the Pentagon to send over an explanation of the command's mission and strategy before he would agree to schedule a confirmation hearing.

While Senator Levin was trying to figure out what Cyber Command was supposed to be protecting and General Alexander was "in the quiet period" before his hearing, I wasn't too clear on what Homeland Security was supposed to protect. Therefore I went to the source and asked Secretary Janet Napolitano. She graciously agreed to meet with me at her department's headquarters. Unlike other cabinet departments, which tend to be headquartered in monumental edifices or modern office blocks near the National Mall, the newest department is run from a barbed-wire-enclosed encampment in northwest Washington, D.C. Behind the wire are a series of low-rise redbrick buildings that, seen from the street, appear like a Nazi army kaserne. It is little wonder that when civil servants were forced to move in they gave the place the nickname Stalag 13, after the fictional German prison camp in the long-running television comedy show *Hogan's Heroes*.

In fact, the facility had been the headquarters of the U.S. Navy's cryptographical service, the predecessor of the new 10th Fleet. Like U.S. Navy bases everywhere, this one came with a little white church and cute little street signs. One street is named "Intelligence Way." To get to the Secretary's office, we walked through a seemingly endless sea of gray Dilbert cubicles. Napolitano's personal office was only slightly better. For the former Governor of Arizona, the dismal ten-by-twelve-foot office was a distinct comedown. Nonetheless, she had managed to cram a bronco-busting saddle into one corner. But the place had a temporary feel to it, six years after the department had been created. "We're moving to a big new headquarters," the Secretary explained, trying to emphasize the positive. The new headquarters, on the grounds of St. Elizabeth's, Washington, D.C.'s shuttered insane asylum, would be ready in year ten of the department's existence, maybe.

"Even though the government was closed for a holiday yesterday, I spent it meeting with executives form the financial sector, talking to them about cyber security," Napolitano began. It was Cyber Security Awareness Month at the department and she had scheduled a number of events. I asked her what the greatest cyber security threat was. "The highly skilled lone hacker, cyber criminal cartels . . ." she replied. Well, what if there were a cyber war, I asked. "The Pentagon would have the lead in a war, but we would do consequence management of any damage in the U.S." What about preventing the damage so that there would be fewer consequences to manage? "We are growing the capability so that we might be able to protect the dot-gov domain.

Well, if U.S. Cyber Command is protecting dot-mil and you will one day protect dot-gov, who is protecting everything else, like the critical infrastructure, which is in the private sector? "We work with the private sector groups, the Information Sharing and Analysis Centers in the eighteen critical industries, to share information with them." That is not the same thing as the U.S. government protecting the critical infrastructure from cyber war attacks, is it? No, the Secretary admitted, it wasn't. Doing that, she suggested, was not Homeland Security's job.

Homeland Security is developing a system to scan cyber traffic going to and from federal departments, looking for malware (viruses, worms, etc.). The immodestly named "Einstein" system had grown from mere traffic flow monitoring (Einstein 1) to intrusion and malware detection (Einstein 2) and will soon attempt to block Internet packets that appear to be malware (Einstein 3). As part of the effort to defend the government sites, Homeland and the General Services Administration are attempting to reduce the number of portals from the Internet to the dot-gov domain. Then Homeland will place Einstein 3 on each of those portals into dot-gov to scan for malware. The Einstein network will be run by Homeland's newly consolidated cyber security division, the National Cybersecurity and Communications Integration Center in Ballston, Virginia.

If DHS can get this to work, I asked, why just limit it to protecting the federal government? "Well, we may want to look later on at taking it out more broadly." Secretary Napolitano, who is a lawyer and a former federal prosecutor, added that there would be legal and privacy hurdles to having the government scanning the public Internet for cyber war attacks. Well, then, could she employ regulatory authority to make critical infrastructure improve their own ability to defend from cyber war attacks, and to regulate the ISPs or the electric power companies? To her credit, Secretary Napolitano did not rule those possibilities out either, even though President Obama himself had seemed to in his cyber security speech in May 2009. But regulation, she noted, would come only after information sharing and voluntary measures had been shown to fail, and in year one of the Obama Administration it was too early to make that judgment. Of course, information sharing and voluntary measure approach had been tried for over a decade.

What was within her responsibilities was to secure the dot-gov domain, and Napolitano was pleased to report that DHS was looking for one thousand new employees with cyber security skills. Immediately critics wondered publicly why highly qualified cyber geeks would want to work for Homeland when everyone from Cyber Command to Lockheed and Bank of America was recruiting them. Napolitano said she was working to get the personnel rules changed so that she could pay salaries competitive with the private sector, and she was looking into creating satellite offices in California and other places away from Washington where geeks "might prefer to live." I thought I heard in her voice the longing for back home that many in the Washington bureaucracy secretly harbor. As we left the Secretary's office, the head of the U.S. Coast Guard, Admiral Thad Allen, was waiting outside. "Glad to see you survived the interview with Dick," the Admiral joked. "I survived," the Secretary replied, "but now I'm depressed about cyber war."

Why had Clinton, Bush, and then Obama failed to deal successfully with the problem posed by America's private-sector vulnerability to cyber war? People who have worked on this issue for years all have slightly different answers, or differences in emphasis. Let's explore six of the reasons they most often give.

1. THE GREATEST TRICK

The first reason you hear is that many cyber attacks that have happened have left behind no marks, no gaping crater like Manhattan's Ground Zero. When private-sector firms have their core intellectual property stolen, they usually don't even know it happened. To understand the problem that creates, imagine that you work in a museum with valuable objects, let's say sculptures and paintings. When you leave the museum at the end of the day, you turn on an alarm system and make sure that the video recorder is running and is connected to the surveillance cameras. In the morning, you return. The alarm has not gone off overnight, but just to be sure, you scan through the video of the last twelve hours and satisfy yourself that no one was inside the museum while you were gone. Finally, you check all the sculptures and paintings to be sure that they are still there. All is well. Why ever would you then think you had a security problem?

That is essentially the situation that the Pentagon was facing in the late 1990s and continues to face today. There may be some low-level activity of people trying to penetrate their networks, but doesn't the security software (firewalls, intrusion-detection systems, intrusion-prevention systems) deal effectively with most of the threats? Why would the brass think that their intellectual property, their crown jewels, war plans, engineering drawings, or software was now residing on hard drives in China, Russia, or anywhere other than just on their systems?

The difference between art thieves and world-class hackers is that with the best of the cyber thieves, you never know you were a victim. "Hell, the U.S. government does [number withheld] penetrations of foreign networks every month," one intelligence official told me. "We never get caught. If we are not getting caught, what aren't we catching when we're guarding our own?" How do you convince someone that they have a problem when there is no evidence you can give them? The data isn't missing like the Vermeer that was snatched from the Isabella Stewart Gardner Museum in Boston in 1990. This sounds like a new problem, unique to cyberspace. Historians of military intelligence, however, have heard this tale before.

In the Cold War the United States Navy was confident that it could defeat the Soviet naval forces if it ever came to a shooting war, until they learned that a family of Americans had given the Soviets a unique advantage. The Walker family, including an employee at the National Security Agency and his son in the U.S. Navy, had supplied the Soviets with the Navy's top-secret codes, the cryptology that scrambled and unscrambled messages to and from our ships. The Red Navy knew where our ships were, where they were going, what they were ordered to do, and which major weapons and other systems onboard were not working. We were unaware that the Soviets knew these things because, although we assumed that they were intercepting our message traffic coming over radio frequencies, we were very confident that they could never unscramble our code. They probably never could have, until they bought the descrambling key from some trusted Americans.

The U.S. Navy's smug arrogance about the security of its Cold War codes was hardly unique in the history of code-breaking: the Japanese thought that no one could read their naval codes during World War II, but the United States and the United Kingdom were doing just that. Some historians believe that the U.S. Navy defeated the Imperial Japanese Navy precisely because of code-breaking skills. Certainly the decisive U.S. victory in the Battle of Midway was due to the advanced knowledge of Japanese plans gained from code-breaking. It is a reasonable assumption that over several decades many nations' codes, presumed to be unbreakable by their users, were (or are) actually being read by others.

Even though historians and national security officials know that there are numerous precedents for institutions thinking their communications are secure when they are not, there is still resistance to believing that it may be happening now, and to us. American military leaders today cannot conceive of the possibility that their Secret (SIPRNET) or Top Secret intranet (JWICS) is compromised, but several experts I spoke to are convinced that it is. Many corporate leaders also believe that the millions of dollars they have spent on computer security systems means they have successfully protected their company's secrets. After all, if anybody had gotten inside their secret files, the intrusion detection system software would have sounded an alarm. Right?

No, not necessarily. And even if the alarm did go off, in many cases that would not have caused anyone to do anything very quickly in response. There are ways of penetrating networks and assuming the role of the network administrator or other authorized user without ever doing anything that would cause an alarm. Moreover, if an alarm does go off, it is often such a routine occurrence on a large network that nothing will happen in response. Perhaps the next day someone will check the logs and notice that a couple of terabytes of information were downloaded and transmitted outside of the network to some compromised server, the first stop on a multistage trip intended to obscure the final destination. Or, perhaps, no one will notice that anything ever happened. The priceless art is still on the museum walls. And if that is the case, why should the government or the bottom-line-conscious executive do anything?

I mentioned in chapter 2 the 2003 phenomenon code-named Titan Rain. Alan Paller, a friend who runs the SANS Institute, a cyber security education and advocacy group, described what happened on one afternoon in that case, November 1, 2003.

At 10:23 p.m. the Titan Rain hackers exploited vulnerabilities at the U.S. Army Information Systems Engineering Command at Fort Huachuca, Arizona.

At 1:19 a.m. they exploited the same hole in computers at the Defense Information Systems Agency in Arlington, Virginia.

At 3:25 a.m. they hit the Naval Ocean Systems Center, a Defense Department installation in San Diego, California.

At 4:46 a.m. they struck the U.S. Army Space and Strategic Defense installation in Huntsville, Alabama.

There were lots of days like that. Not only were Defense facilities hit, but terabytes of sensitive information left NASA labs, as well as the computers of corporations such as Lockheed Martin and Northrop Grumman, which have been given contracts worth billions of dollars to manage security for DoD networks. Cyber security staffs tried to figure out the techniques being used to penetrate the networks. And their blocking efforts seemed to work. One participant in these defensive efforts told us that "Everyone was all self-congratulatory." He shook his head, pulled a grimace, and added softly, ". . . till they realized that the attacker had just gone all stealthy, but was probably still stealing us blind. We just couldn't see it anymore." The case names Moonlight Maze and Titan Rain are now best thought of as fleeting glimpses of a much broader campaign, most of which went unseen. It may seem somewhat incredible that terabytes of information can be removed from a company's network without that company being able to stop it all from going out the door. In the major cases we know about, the companies or federal organizations usually did not even detect that an exfiltration of data had occurred until well after it had taken place. All of these victims had intrusion-detection systems that are supposed to alarm when an unauthorized intruder attempts to get on a network. Some sites even had the more advanced intrusion-*prevention* systems, which not only alarm but also automatically take steps to block an intruder. The alarms remained silent. If you have a mental image of every interesting lab, company, and research facility in the U.S. being systematically vacuum cleaned by some foreign entity, you've got it right. That is what has been going on. Much of our intellectual property as a nation has been copied and sent overseas. Our best hope is that whoever is doing this does not have enough analysts to go through it all and find the gems, but that is a faint hope, particularly if the country behind the hacks has, say, a billion people in it.

One bright spot in this overall picture of data going out the door unhindered is what happened at Johns Hopkins University's Advanced Physics Laboratory (APL), outside Baltimore. APL does hundreds of millions of dollars of research every year for the U.S. government, from outer-space technology to biomedicine to secret "national security" projects. APL did discover in 2009 that it had huge amounts of data being secretly exfiltrated off its network and they stopped it. What is very telling is the way in which they stopped it. APL is one of the places that is really expert in cyber security and has contracts with the National Security Agency. So one might think that they were able to get their intrusion systems to block the data theft. No. The only way in which these cyber experts were able to prevent their network from being pillaged was to disconnect the organization from the Internet. APL pulled the plug and isolated its entire network, making it an island in cyberspace. For weeks, APL's experts went throughout the network, machine by machine, attempting to discover trapdoors and other malware. So the state of the art in really insuring that your data does not get copied right off your network appears to be to make sure that you are in no way connected to anybody. Even that turns out to be harder than it may seem. In large organizations, people innocently make connections to their home computers, to laptops with wi-fi connections, to devices like photocopiers that have their own connectivity through the Internet. If you are connected to the Internet in any way, it seems, your data is already gone.

The really good cyber hackers, including the best government teams from countries such as the U.S. and Russia, are seldom stumped when trying to penetrate a network, even if its operators think the network is not connected in any way to the public Internet. Furthermore, the varsity teams do something that causes network defenders to sound like paranoids. They never leave any marks that they were there, except when they want you to know. Think of Kevin Spacey's character's line in the movie *The Usual Suspects*: "The greatest trick the devil ever pulled was convincing the world he didn't exist."

2. VEGAS, BABY

Another reason given for why there has not been a groundswell sufficient to address America's vulnerability to cyber war is that the "thought leadership" group in the field can't agree on what to do. To test that hypothesis, I went in search of the "thought leaders" in what you might think was one of the more unlikely places to find them, Caesars Palace, in Las Vegas, in the 104-degree heat of August 2009.

Caesars is an incongruous site on any day, filled as it is with statues and symbols of an empire that fell over fifteen centuries ago scattered among blinking slot machines and blackjack tables. At Caesars, conference rooms with names like the Colosseum and the Palatine are not crumbling ruins, but are state-of-the-art meeting facilities, with white boards, flat screens, and flashing control consoles. Every summer for the past dozen years, when the more mainstream conventions wrap up the Vegas conference season and the room prices drop, a slightly different kind of crowd descends on the Strip. They are mainly men, usually in shorts and T-shirts, often with backpacks, BlackBerrys, and Mac laptops. Few of them drop into the fashion-forward Hugo Boss, Zegna, or Hermès shops in Caesars Forum, but they have almost all been to the *Star Trek* show over at the Hilton. The crowd are hackers, and in 2009 over four thousand of them showed up for the Black Hat conference, enough information technology skill in one place to wage cyber war on a massive scale.

Despite the name, Black Hat is actually now a gathering of "white hat," or "ethical," hackers, people who are or work for chief information officers (CIOs) or chief information security officers (CISOs) at banks, pharmaceutical firms, universities, government agencies, almost every imaginable kind of large (and many medium-sized) company. The name Black Hat derives from the fact that the highlights of the show every year are announcements by hackers that they've figured out new ways to make popular software applications do things they were not designed to do. The software companies used to think of the conference as a meeting of bad guys. Usually the demonstrations show that software's writers were not sufficiently security conscious, with the result that there is a way to penetrate a computer network without authorization, maybe even gain control of a network.

Microsoft was the butt of the conference's hacking for years, and the executives in Redmond looked forward annually to Black Hat the way most of us anticipate a tax audit. In 2009 the attention turned to Apple, because of the increasing popularity of its products. The most-discussed demonstration concerned how to hack an iPhone with a simple SMS text message. As much as Bill Gates, or now maybe even Steve Jobs, might like it to be illegal for people to find and publicize the flaws in their products, it is not a crime to do so. A crime occurs only when a hacker uses the method he's developed (the "exploit") to utilize the flaw he's discovered in the software (the "vulnerability") to get him into a corporate or government network ("the target") where he is not authorized to be. Of course, once a vulnerability is publicized at Black Hat, or, worse yet, once an exploit is disseminated, anyone can attack any network running the flawed software.

I got into a little bit of trouble in 2002 for suggesting in my Black Hat keynote address that it was a good thing that hackers were discovering flaws in software. I was Special Advisor for Cyber Security to President Bush at the time. Someone, presumably in Redmond, thought it wrong for a nice conservative Republican White House to be encouraging illegal acts. Of course, what I actually said was that when the ethical hackers discovered flaws they should first tell the software maker, and then, if they got no response, call the government. Only if the software maker refused to fix the problem, I said, should the hackers go public. My logic was that if the hackers at Black Hat could discover the software flaws, China, Russia, and others probably could, too. Since those engaged in espionage and crime would find out anyway, it was better if everyone else knew.

Public knowledge of a "bug" in software would probably mean two things: (1) most sensitive networks would stop using the software until it got fixed, and (2) the software manufacturer would be shamed into fixing it, or pressured to do so by its paying customers, such as banks and the Pentagon.

Comments like that did not endear me to certain corporate interests. They also didn't like it when, again in 2002, I was the keynote speaker at the annual RSA conference. The RSA conference is a gathering of about 12,000 cyber security practitioners. It is an occasion for many late-night parties. My keynote was early in the morning. I was standing backstage, thinking about how I needed more coffee. The band Kansas had been brought in and was playing loudly in the big hall. When they were done, I was supposed to walk out onstage through a cloud of theatrical smoke. You get the picture. Thinking of my caffeine needs, I noted shortly after starting the speech that a recent survey had shown that many large companies spent more money on free coffee for their employees and guests than they did on cyber security. To which I added, "If you're a big company and spend more on coffee than on cyber security, you *will* be hacked." Pause. Then go for it. "What's more, if those are your priorities, you *deserve* to be hacked." Dozens of irate telephone calls from corporate officials followed.

RSA is very corporate. Black hat is a lot more fun. The thrill at Black Hat is going into a dimly lit ballroom and seeing someone unaccustomed to public speaking projecting lines of code on a presentation screen. Hotel staff servicing the conference always look quizzical when a meeting room erupts in laughter or applause, which they do a lot, because to the average person nothing is being said that is obviously humorous, praiseworthy, or for that matter even understandable. Perhaps the only thing that most Americans would generally follow if they wandered off course into Black Hat while looking for the roulette tables is the conference's Hacker

Court, mock trials with judges who seek to establish what sort of hacking should really be considered unethical. Apparently hacking the hackers is not in that category. Most conferencegoers just accept that they should have their wi-fi applications turned off on their laptops. Signs throughout the vast conference area note that the wi-fi network should be considered "a hostile environment." The warning is about as necessary as a placard at an aquarium noting that there is no lifeguard on duty in the shark tank.

In 2009, conference organizer Jeff Moss broke with tradition by scheduling one meeting at Black Hat Vegas that was not open to all attendees. Indeed, Moss, who dressed only in black during the conference, limited the attendance at that meeting to thirty people, instead of the usual 500 to 800 who crowd each of the six simultaneous sessions that take place five or more times a day during the conference. The invitation-only session was populated by a group of "old hands," people who knew where the virtual bodies were buried in cyberspace: former government officials, current bureaucrats, chief security officers in major corporations, academics, and senior IT company officials. Moss's question to them: What do we want the new Obama Administration to do to secure cyberspace? In a somewhat unorthodox move, the Obama Administration had placed Moss on the Homeland Security Advisory Board, so there was some chance that his reporting of the group's consensus views would be heard, assuming the group could strike a consensus.

To their surprise, the group reached general accord on a few things, as well as polarized disagreement on others. Where the consensus emerged was around five points. First, the group was all in favor of returning to the days when the federal government spent a lot on cyber security research and development. The agency that had done so, and which had also funded the creation of the Internet, DARPA (the Defense Advanced Research Projects Agency), had essentially abandoned the Internet security field during the Bush (43)

Administration and instead focused attention on "netcentric war-fare," apparently oblivious that such combat depended upon cyber-space being secure.

Second, there was a slight majority in favor of "smart regulation" of some aspects of cyber security, like maybe federal guidelines for the Internet backbone carriers. The smart part was the idea of gov-ernment regulators specifying goals, rather than micromanaging by dictating means. Most thought, however, that the well-entrenched interest groups in Washington would successfully lobby Congress to block any regulation in this area. Third, the group thought worry-ing about who did cyber attacks, the so-called attribution problem, was fruitless and that people should instead focus on "resilience." Resilience is the concept that accepts that a disruptive or even de-structive attack will occur and advocates planning in advance for how to recover from such devastation.

The fourth consensus observation was that there really should be no connectivity between utility networks and the Internet. The idea of separating "critical infrastructure" from the open-to-anyone Internet seemed pretty obvious to the seasoned group of informa-tion security specialists. In a ballroom down the hall, however, the Obama Administration's ideas about a Smart Electric Grid were being flayed by several hundred other security specialists, precisely because the plans would make the electric power grid, that sine qua non for all the other infrastructure, even more vulnerable to unau-thorized penetration and disruption from the anonymous creatures who prowl the Internet.

The final point on which the "wise men" (including three women) were able to generally agree was that nothing would happen to solve the woes of cyberspace security until someone showed what is so lacking now: leadership. This observation apparently did not seem ironic to the group, who, arguably, were the leaders of the elite information technology security specialists in the country. Yet they looked to the Obama Administration for leadership in the area. At that point the Obama White House had already called over thirty people to see if they were interested in being the administration's leader on cyberspace security. The search went on in Washington, as did the demonstrations down the hall of how to hack systems. As the "thought leaders" wandered out of the Pompeii Room somewhat dejected, hoping for leadership, they could hear erupting, probably from the Vesuvius Room, the sound of hundreds wailing as a hacker virtually sliced apart another iPhone. We did not rush over to see what application had been hacked. Instead, we went off to the blackjack tables, where the odds of our losing seemed less than those for American companies and government agencies hoping to stay safe in cyberspace.

3. PRIVACY AND THE R WORD

When both the left and the right disagree with your solution to a problem, you know two things: (1) you are probably on the correct path, and (2) you stand almost no chance of getting your solution adopted. Many of the things that have to be done to reduce America's vulnerability to cyber war are anathema to one or the other end of the political spectrum. That is why they have not been embraced thus far.

I will discuss the details of what might be done in the next chapter, but I can tell you now that some of the ideas will require regulation and some will have the potential, if abused, to violate privacy. In Washington, one might as well advocate random forced abortions as suggest new regulation or create any greater privacy risks.

My position on regulation is that it is neither good nor bad inherently; it depends upon what the regulation says. Complex, 1960s-style federal regulations generally serve only the Washington law firms where they are written, and where policies to avoid them are devised at $1,000 an hour. "Smart regulation," as discussed at Black Hat, articulates an end state and allows the regulated to figure out how best to get to it. Regulation that puts a U.S. company at an economic disadvantage to a foreign competitor is usually unwise, but a regulatory even playing field that passes on minimal costs to users does not seem to me to be one of the works of Satan. Regulations where compliance is not audited or enforced are worthless, almost as troubling as regulations requiring the hovering presence of federal officials. Third-party audits and remote compliance verification generally seem like sensible approaches. Refusal to regulate, or audit, or enforce, often results in things like the 2008 market crash and recession, or lead paint in children's toys. Overregulation sometimes creates artificially high consumer prices and requirements that do little or nothing to solve the original problem, and suppresses creativity and innovation.

On privacy rights, and civil liberties in general, I am far more categorical. We need to be vigilant, lest government erode our rights. This is not an unjustified fear. Well-meaning provisions of the Patriot Act were abused in recent years. Other restrictions on government action, including those in the Bill of Rights and in the Foreign Intelligence Surveillance Act, were simply ignored. If what we need to do to defend ourselves from cyber war opens the possibility of further government abuse, we will need to do more than simply pass laws making such government action illegal. That has not stopped some in the past. (Cheney, I'm thinking of you here.) We will also have to create empowered, independent organizations to investigate whether abuses are occurring and to bring legal action against those who are violating privacy laws and civil liberties. The safest way to deal with the threat of further abuse is, of course, not to create new programs that government officials could misuse to violate our rights. There may be times, however, as in the case of cyber war, when we should examine whether effective safeguards can be put in place so that we can start new programs that entail some risk.

4. CASSANDRAS AND RED HERRINGS

Part of the reason that we are so unprepared today is the "boy who cried wolf too soon" phenomenon. Sometimes the boy who cries wolf can see the wolf coming from a lot farther away than everyone else. The Joint Security Commission of 1994, the Marsh Commission of 1997, the Center for Strategic and International Studies (CSIS) commission of 2008, the National Academy of Science commission of 2009, and many more in between have all spoken of a major cyber security or cyber war risk. They have been criticized by many as Cassandras, the type of people who are always predicting disaster. The earth will be hit by a giant meteor. A shift in magnetic north from one pole to another will cause solar wind that will destroy the atmosphere. Well, almost all real experts in the relevant fields of science believe the meteor and pole-shifting scenarios will happen. They just do not know when, and therefore we probably should not get too excited. The various commissions and groups warning about cyber war have not really been wrong about the timing; they were warning us when we had sufficient time to do something in advance of a disaster. It is worth remembering that, despite the bad rap she gets, Cassandra was not wrong about her predictions; she was simply cursed by Apollo never to be believed.

Unfortunately, one thing that is too often believed is that there is a threat from "cyber terrorism." Cyber terrorism is largely a red herring and, in general, the two words "cyber" and "terrorism" should not be used in conjunction because they conjure up images of bin Laden waging cyber war from his cave. He probably can't, at least not yet. (Moreover, he's probably not in a cave, more likely a cushy villa.) Indeed, we do not have any good evidence that terrorists have ever staged cyber war attacks on infrastructure.

To date, terrorists haven't so much attacked the Internet or used the Internet to attack physical systems as they have used it to plan and coordinate attacks on embassies, railroads, and hotels. They have also used the Internet to raise funds, recruit, and train. After al Qaeda lost their training grounds in Afghanistan following 9/11, much of what went on there shifted to the web. Training videos on how to build improvised explosive devices or how to stage beheadings were just as effective delivered over a remote learning system as they were at a remote training camp. The web kept terrorists from having to travel for training, which used to be a very good opportunity for international law enforcement to catch a would-be terrorist. Remote training also kept a bunch of terrorists from congregating in one place long enough for a cruise missile strike. While Internet training has been a huge danger, spawning "lone wolf" attacks by terrorists who never had any connection to al Qaeda central, what al Qaeda and other groups really excel at is using the Internet for propaganda. Producing videos of beheadings and spreading radical interpretations of the Koran across the Internet has allowed terrorist groups to reach a wide audience and to do so with relative anonymity.

While al Qaeda has thus far not been capable of staging a cyber attack, that could very well change. As with any developing technology, the cost and other barriers to entry are going down each year. Staging a devastating cyber attack would not require a major industrial effort like building a nuclear bomb. Understanding the control software for an electric grid, however, is not a widely available skill. It is one thing to find a way to hack into a network, and quite another to know what to do once you're inside. A well-funded terrorist group might find a highly skilled hacker club that would do a cyber attack in return for a lot of money, but that has not happened to date. One of the reasons for that may simply be that most hackers think that al Qaeda members are crazy, dangerous, and untrustworthy. When criminal hacker groups think of others that way, you know the real terrorists are pretty far out there.

5. MONEY TALKS

Another reason for inertia is that some people like things the way they are. Some of those people have bought themselves access. I mentioned earlier that George W. Bush's first reaction when told of a possible cyber security crisis was to ask what a certain computer industry CEO who was one of his biggest campaign donors thought about it. You had probably already guessed that the Bush Administration was not interested in playing hardball with the private sector. The first *Homeland Security Strategy of the United States*, put out in 2003, reads like a conservative economic textbook on the power of the free market. You may be surprised, however, at how Democratic administrations have also been captured by these arguments. You might think that the new Democratic administration would be in favor of finally solving the market failure on cyber security by introducing some new regulation, but you would be wrong. To understand why, let's go to a party.

It was a lavish affair. All the big names in Washington were there. Over 250 guests joined to celebrate the marriage of Melody Barnes to Marland Buckner. Barnes, President Obama's domestic policy advisor, had known her husband-to-be for years before they started dating; their acquaintance goes back to her time on Capitol Hill, working for Ted Kennedy, and to his as Chief of Staff to Harold Ford, Jr., of Tennessee. After a short ceremony at the People's Congregational United Church of Christ, the newlyweds and their guests retired to Washington's Mellon Auditorium, which had been converted into a "South Beach–style" lounge, with hints of silver and a floral theme for each table that was heavy on orchids. The locally sourced, carbon-neutral menu featured short ribs, sea bass, and a selection of spring vegetables elegantly arranged in bento boxes, followed by sliders and fries to keep the guests' energy up until they were released at some point after midnight.

What the *New York Times* Weddings and Celebrations reporter described as "a bevy of Obama Administration officials" in attendance included White House Chief of Staff Rahm Emanuel and Valerie Jarrett, a White House senior advisor and Assistant to the President for Intergovernmental Relations. My friend Mona Sutphen, Deputy Chief of Staff, danced the night away, as did former Clinton Chief of Staff John Podesta. Also in attendance, but not noted by the *Times*, were a bevy of Microsoft executives. Buckner, a former director of government affairs at the world's largest software company and now an independent registered lobbyist, had also invited some friends. Since going out on his own in 2008, Buckner took in lobbying fees, more than a third of which were from Microsoft. It is too bad *Mother Jones* doesn't do weddings. Their reporter might have noted that on that night, the Obama Administration was, quite literally, in bed with Microsoft.

Microsoft makes OpenSecret.org's top 30 list of "Heavy Hitters," donating to political causes. While most of the organizations on that list are trade associations, Microsoft is one of only seven companies that make the cut. Of course, Microsoft was making up for lost time. Before the company's battle with the Justice Department over antitrust issues in the late 1990s, the West Coast–based company wanted nothing more than to be left alone and stayed out of politics. Before 1998, Microsoft and its employees were little inclined to spend their stock options supporting East Coast politicians. That all changed when Clinton Administration lawyers argued that the marketing of Windows was intended to create a monopoly. Donations started pouring in from newly established political action committees and Microsoft employees alike. And in the years 1998 through 2002, the majority of that money went to Republicans. Then, in 2004, maybe disgusted by the war or maybe misunderestimating the Bush campaign, Microsoft began donating to Democrats at almost twice the rate than to Republicans. In 2008, Microsoft beat those numbers, giving $2.3 million to Democrats and only $900,000 to Republicans.

Maybe Microsoft's PACs and employees have good intentions, like so many Americans who donated money and time to the Obama campaign who wanted nothing more than to see Obama in office. Marland Buckner told a reporter for Media General News Service that he would "follow White House rules 'to the letter' to avoid any conflict of interest due to Barnes's new job, and promised not to use his relationship with his spouse to attract clients. But Microsoft the corporation has an agenda that is very clear: don't regulate security in the software industry, don't let the Pentagon stop using our software no matter how many security flaws it has, and don't say anything about software production overseas or deals with China.

Microsoft has vast resources, literally billions of dollars in cash, or liquid asset reserves. Microsoft is an incredibly successful empire built on the premise of market dominance with low-quality goods. For years, Microsoft's operating system and applications, like its ubiquitous Internet browser, have been prepackaged on the computers we buy. Getting an alternative was a time-consuming and problematic task, until Apple began to open stores and advertise in the last decade.

To be fair, Microsoft did not originally intend its software to be running critical systems. Therefore, its goal was to get the product out the door fast and at a low cost of production. It did not originally see any point to investing in the kind of rigorous quality assurance and quality control processes that NASA insisted on for the software used in human space-flight systems. The problem is that people did start using Microsoft products in critical systems, from military weapons platforms to core banking and finance networks. They were, after all, much cheaper than custom-built applications.

Every once in a while there is a wave of government efficiency improvements that brings federal government agencies up to date with the cost-saving approaches being used in industry. One of them was called the COTS campaign. The idea was to use commercial off-the-shelf (COTS) software to replace specialized software that in the past the government would have ordered up. Throughout the Cold War, the Pentagon had led much of this country's technological innovation. I remember being told that there were cameras without film that had been developed for the government. (I could not quite understand how that would work—until I bought one at Best Buy a decade later.) Only after military applications were developed did the technology eventually leak out for commercial use.

COTS stood that process on its head. Before the 1990s, most of the Pentagon's software applications were purpose-built in-house or by a small number of trusted defense contractors. No two systems were alike, which was how the defense contractors wanted it. The systems they built were extremely expensive. They also made it very difficult for defense systems to work interoperably. The COTS movement reduced the costs and allowed the Pentagon to create interoperable systems because they all used the same computer languages and the same operating systems. More and more applications were developed. Sensor grids were netted together. The 5.5-million-computer Global Information Grid, or GIG, was created. Netcentric warfare provided a huge advantage for the U.S. military, but it also introduced a huge vulnerability.

COTS brought to the Pentagon all the same bugs and vulnerabilities that exist on your own computer. In 1997, the U.S. Navy found out just how dangerous it could be to rely on these systems for combat operations. The USS *Yorktown*, a Ticonderoga-class cruiser, was retrofitted as the test bed for the Navy's "smart ship" program. The *Yorktown* had been outfitted with a network of twenty-seven Pentium-powered workstations all running Windows NT, all tied to a Windows server. The system controlled every aspect of ship operations, from bridge operations to fire control to engine speed. When the Windows system crashed, as Windows often does, the cruiser became a floating i-brick, dead in the water.

In response to the *Yorktown* incident and a legion of other failures of Windows-based systems, the Pentagon began to look at Unix and the related Linux systems for critical operations. Linux is an open-source system. What that means is that the computer code for the operating system can be viewed and edited by the user. With Windows (and most other commercial software), the source code is considered to be proprietary and is heavily guarded. Open source had a number of advantages for the Pentagon. First, Pentagon programmers and defense contractors could customize the software to make it operate the way they wanted. They could slice and dice the code to eliminate parts of the operating system that they did not need and that could introduce bugs into the system. Second, after reducing the size of the operating system, they could then run what the software industry refers to as "tools" on the remaining lines of code to try and identify bugs, malicious code, and other vulnerabilities.

Microsoft went on the warpath against Linux to slow the adoption of it by government agencies, complete with appearances before congressional committees, including by Bill Gates. Nonetheless, because there were government agencies using Linux, I asked NSA to do an assessment of it. In a move that startled the open-source community, NSA joined that community by publicly offering fixes to the Linux operating system that would improve its security. Microsoft gave me the very clear impression that if the U.S. government promoted Linux, Microsoft would stop cooperating with the U.S. government. While that did not faze me, it may have had an effect on others. Microsoft's software is still being bought by most federal agencies, even though Linux is free.

The banking and finance industry also started to look at open-source alternatives after the repeated failure of Microsoft systems had cost the finance industry hundreds of millions a year. In 2004, a banking industry group, the Financial Services Roundtable, sent a delegation of computer security specialists from the banks to Redmond, Washington, to confront Microsoft. They demanded access to the secret Microsoft code. They were denied. They demanded to see the quality-assurance standards Microsoft used so that they could compare them with other software companies. They were denied. Microsoft's position with the U.S. banks is in contrast to the program the company had announced in 2003 whereby, pursuant to agreement, Microsoft provide participating national and international bodies access to its Windows source code, a move designed to address concerns about the security of its operating system. Russia, China, NATO, and the United Kingdom were early participants.

The banks threatened to start using Linux. Microsoft told them the conversion to Linux would be very expensive for them. Moreover, the next version of Windows was being developed under the code name Longhorn. Longhorn would be much better. Longhorn became Vista. Vista went to market later than expected, delayed by flaws discovered in Microsoft's expanded tests program. When Vista was sold, many corporate users experienced problems. Word spread and many companies decided not to buy the new system. Microsoft suggested that it would stop providing support for some of its older systems, forcing customers to upgrade.

Microsoft insiders have admitted to me that the company really did not take security seriously, even when they were being embarrassed by frequent highly publicized hacks. Why should they? There was no real alternative to its software, and they were swimming in money from their profits. When Linux appeared, and later when

Apple started to compete directly, Microsoft did take steps to improve its quality. What they did first, however, was to employ a lot of spokesmen to go to conferences, to customers, and to government agencies lobbying against moves to force improvements in security. Microsoft can buy a lot of spokesmen and lobbyists for a fraction of the cost of creating more secure systems. They are one of several dominant companies in the cyber industry for whom life is good right now and change may be bad.

6. NO, I THOUGHT YOU WERE DOING IT

Change, however, is coming. Like the United States, more and more nations are establishing offensive cyber war organizations. U.S. Cyber Command also has a defensive mission, to defend the Department of Defense. Who defends the rest?

As it stands now, the Department of Homeland Security defends the non-DoD part of the federal government. The rest of us are on our own. There is *no* federal agency that has the mission to defend the banking system, the transportation networks, or the power grid from cyber attack. Cyber Command and DHS think that by defending their government customers they may coincidentally help the private sector a little, maybe. The government thinks it is the responsibility of individual corporations to defend themselves from cyber war. Government officials will tell you that the private sector wants it that way, wants to keep the government out of their systems. After all, they are right that no one in government would know how to run a big bank's networks, or a railroad's, or a power grid's.

When you talk to CEOs and the other C-level types in big companies (chief operating officers, chief security officers, chief information officers, chief information security officers), they all say pretty much the same things: we will spend enough on computer security to protect against the day-to-day threat of cyber crime. We cannot, they say, be expected to know how to, or spend the money to, defend against a nation-state attack in a cyber war. Then they usually add words to the effect of, "Defending against other nations' militaries is the government's job, it's what we pay taxes for."

At the beginning of the era of strategic nuclear war capability, the U.S. deployed thousands of air defense fighter aircraft and ground-based missiles to defend the population and the industrial base, not just to protect military facilities. Every major city was ringed with Nike missile bases to shoot down Soviet bombers. At the beginning of the age of cyber war, the U.S. government is telling the population and industry to defend themselves. As one friend of mine asked, "Can you imagine if in 1958 the Pentagon told U.S. Steel and General Motors to go buy their own Nike missiles to protect themselves? That's in effect what the Obama Administration is saying to industry today."

On this fundamental issue of whose job it is to defend America's infrastructure in a cyber war, the government and industry are talking past each other. As a result, no one is defending the likely targets in a cyber war, at least not in the U.S. In other countries, some of whom might be cyber war adversaries someday, the defense part of cyber war might be doing a little better than it is here.

THE CYBER WAR GAP

We noted earlier that the U.S. may have the most sophisticated and complex cyber war capability, followed soon thereafter by Russia. China and perhaps France are in a close second tier, but over twenty nations have some capability, including Iran and North Korea. Whether or not this ranking is accurate, it is widely believed by cyber warriors. So, one can almost imagine the American geek fighters sitting around after work in some secure location drinking their Red Bulls and chanting "U-S-A, U-S-A," as at the Olympics, or "We're Number One!" as at a high school football game. (My high school was so nerdy we chanted "Sumus Primi!") But are we really number one? That obviously depends upon what criteria you employ.

In cyber offensive capability, the United States probably would rank first if you could develop an appropriate contest. But there is more to cyber war than cyber offensive. There is also cyber dependence, the degree to which a nation relies upon cyber-controlled systems. In a two-way cyber war, that matters. As I discovered when I asked for a cyber war plan to go after Afghanistan in 2001, there are sometimes no targets for cyber warriors. In a two-way cyber war, that gives Afghanistan an advantage of sorts. If they had any offensive cyber capability (they didn't), the cyber war balance would have shifted in an interesting way. There is also the issue of whether a nation can defend itself from cyber war. Obviously, Afghanistan can protect itself just by being there and having no networks, but theoretically a nation may have networks and, unlike us, be able to protect them. Cyber defense capability is also, therefore, a criterion: Can a nation shut off its cyber connectivity to the rest of the world, or spot cyber attacks coming from inside its geographical boundaries and stop them?

While the United States very likely possesses the most sophisticated offensive cyber war capabilities, that offensive prowess cannot make up for the weaknesses in our defensive position. As former Admiral McConnell has noted, "Because we are the most developed technologically—we have the most bandwidth running through our society and are more dependent on that bandwidth—we are the most vulnerable." We have connected more of our economy to the Internet than any other nation. Of the eighteen civilian infrastructure sectors identified as critical by the Department of Homeland Security, all have grown reliant on the Internet to carry out their basic functions, and all are vulnerable to cyber attacks by nation-state actors. Contrast this with China. While China has been developing its offensive cyber capability, it has also focused on defense. The PLA's cyber warriors are tasked with both offense and defense in cyberspace, and unlike in the case of the U.S. military, when they say defense, they mean defense of the nation, not just defense of the military's networks. While I do not advocate an expanded role for the Pentagon in protecting civilian systems in the U.S., there is no other agency or arm of the federal government that has taken on that responsibility. In light of the eschewing of regulation that began in the Clinton Administration and has continued through the Bush Administration and into the Obama Administration, the private sector has not been required to improve security, nor has the government stepped in to actively take on the role. In China, the networks that make up the Chinese Internet infrastructure are all controlled by the government through direct ownership or very close partnership with the private sector. There are no debates about the cost of security when Chinese authorities demand new security measures. The networks are largely segmented between government, academic, and commercial use. The Chinese government has both the power and the means to disconnect China's slice of the Internet from the rest of the world, which it may very well do in the event of a conflict with the United States. The U.S. government has no such authority or capability. In the U.S., the Federal Communications Commission has the legal power to regulate but it largely chosen not to do that. In China, the government can set and enforce standards, but it also goes many steps further.

The "Internet" in China is more like the internal network of a company, an intranet. The government is the service provider and therefore in charge of the network's defense. In China, the government is actively defending the network. Not so in the United States. In the U.S., the government's role is at least one step removed. As mentioned briefly in chapter 2, China's much-discussed Internet censorship, including "the Great Firewall of China," can also provide security advantages. The technology that the Chinese use to screen e-mails for speech deemed illegal can also provide the infrastructure to stop malware. China has also invested in developing its own proprietary operating system that would not be susceptible to existing network attacks, though technical problems have delayed its implementation. China launched and then temporarily halted an effort to install software on all computers in China, software allegedly meant to keep children from gaining access to pornography. The real intent, most experts believe, was to give China control over every desktop in the country. (When word of the plan got out in the hacker community, they quickly found vulnerabilities that could have given almost anyone control over the system, and the Chinese promptly delayed the program.) These efforts show how seriously the Chinese take their defense, as well as the direction their efforts are headed. China, meanwhile, remains behind the United States in the automation of its critical systems. Its electric power system, for example, relies on control systems that require a large degree of manual control. This is an advantage in cyber war.

MEASURING CYBER WAR STRENGTH

It would be great if the only thing we had to take into account in measuring our cyber war strength was one factor, our ability to attack other nations. If that were the only consideration, the United States might do really well when compared to other nations. Unfortunately for us, a realistic measurement of cyber war strength also needs to include an assessment of two other factors: defense and dependence. "Defense" is a measure of a nation's ability to take actions that under attack, actions which will block or mitigate the attack. "Dependence" is the extent to which a nation is wired, reliant upon networks and systems that could be vulnerable in the event of cyber war attack.

To illustrate how these three factors (offense, defense, and dependence) interact, I have created a chart. The chart assigns scores to several countries for each of the three factors. Quibblers will argue with the overly simplistic methodology: I gave each of the three measures equal weight and then added the three scores together to get an overall score for a nation. The scores assigned to each nation are based on my assessment of their offense power, their defensive capability, and the extent to which they are dependent on cyber systems. There is one counterintuitive aspect to the chart: the less wired a nation is, the higher its score on the dependence ranking. Being a wired nation is generally a good thing, but not when you are measuring its ability to withstand cyber war.

OVERALL CYBER WAR STRENGTH

Nation	Cyber Offense	Cyber Dependence	Cyber Defense	Total
U.S.	8	2	1	11
Russia	7	5	4	16
China	5	4	6	15
Iran	4	5	3	12
North Korea	2	9	7	18

The results are revelatory. China has a high "defense" score, in part because it has plans and capability to disconnect the entire nation's networks from the rest of cyberspace. The U.S., by contrast, has neither the plans nor the capability to do that because the cyber connections into the U.S. are privately owned and operated. China can limit cyberspace utilization in a crisis by disconnecting nonessential users. The U.S. cannot. North Korea gets a high score for both "defense" and "lack of dependence." North Korea can sever its limited connection to cyberspace even more easily and effectively than China can. Moreover, North Korea has so few systems dependent upon cyberspace that a major cyber war attack on North Korea would cause almost no damage. Remember that cyber dependence is not about the percentage of homes with broadband or the per capita number of smart phones; it's about the extent to which critical infrastructures (electric power, rails, pipelines, supply chains) are dependent upon networked systems and have no real backup.

When you think about "defense" capability and "lack of dependence" together, many nations score far better than the U.S. Their ability to survive a cyber war, with lower costs, compared to what would happen to the U.S., creates a "cyber war gap." They can use cyber war against us and do great damage, while at the same time they may be able to withstand a U.S. cyber war response. The existence of that "cyber war gap" may tempt some nation to attack the United States. Closing that gap should be the highest priority of U.S. cyber warriors. Improving our offensive capability does not close the gap. It is impossible to reduce our dependence on networked systems at this point. Hence, the only way we can close the gap, the only way we can improve our overall Cyber War Strength score, is to improve our defenses. Let's take a look at how we might do that.

CHAPTER FIVE

TOWARD A DEFENSIVE STRATEGY

Military theorists and statesmen, from Sun Tzu to von Clausewitz to Herman Kahn, have for centuries defined and redefined military strategy in varying ways, but they tend to agree that it involves an articulation of goals, means (broadly defined), limits (perhaps), and possibly sequencing. In short, military strategy is an integrated theory about what we want do and how, in general, we plan to do it. In part because Congress has required it, successive U.S. administrations have periodically published a National Security Strategy and a National Military Strategy for all the world to read. Within the military, the U.S. has many substrategies, such as a naval strategy, a counterinsurgency strategy, and a strategic nuclear strategy. The U.S. government has also publicly published strategies for dealing with issues wherein the military plays only a limited role, such as controlling illegal narcotics trafficking, countering terrorism, and stopping the proliferation of weapons of mass destruction. Oh yes, there is also that National Strategy to Secure Cyberspace dating back to 2003; but there is no publicly available cyber war strategy.

In the absence of a strategy for cyber war, we do not have an integrated theory about how to address key issues. To prove that, let's play Twenty Questions and see if there are agreed-upon answers to some pretty obvious questions about how to conduct cyber war:

- What do we do if we wake up one day and find the western half of the U.S. without electrical power as the result of a cyber attack?
- Is the advent of cyber war a good thing, or does it place us at a disadvantage?
- Do we envision the use of cyber war weapons only in response to the use of cyber war weapons against us?
- Are cyber weapons something that we will employ routinely in both small and large conflicts? Will we use them early in a conflict because they give us a unique advantage in seeking our goals, such as maybe effecting a rapid end to the conflict?
- Do we think we want to have plans and capabilities to conduct "stand-alone" cyber war against another nation? And will we fight in cyberspace even when we're not shooting at the other side in physical space?
- Do we see cyberspace as another domain (like the sea, airspace, or outer space) in which we must be militarily dominant and in which we will engage an opponent while simultaneously conducting operations in other domains?
- How surely do we have to identify who attacked us in cyberspace before we respond? What standards will we use for these identifications?

- Will we ever hide the fact that it was us who attacked with cyber weapons?
- Should we be hacking into other nations' networks in peacetime? If so, should there be any constraints on what we would do in peacetime?
- What do we do if we find that other nations have hacked into our networks in peacetime? What if they left behind logic bombs in our infrastructure networks?
- Do we intend to use cyber weapons primarily or initially against military targets only? How do we define military targets?
- Or do we see the utility of cyber weapons being their ability to inflict disruption on the economic infrastructure or the society at large?
- What is the importance of avoiding collateral damage with our cyber weapons? How might avoiding it limit our use of the weapons?
- If we are attacked with cyber weapons, under what circumstances would, or should, we respond with kinetic weapons? How much of the answer to this question should be publicly known in advance?
- What kind of goals specific to the employment of cyber weapons would we want to achieve if we conducted cyber war, either in conjunction with kinetic war or as a stand-alone activity?
- Should the line between peace and cyber war be brightly delineated, or is there an advantage to us in blurring that distinction?
- Would we fight cyber war in a coalition with other nations, helping to defend their cyberspace and sharing our cyber weapons, tactics, and targets?

- What level of command authority should authorize the use of cyber weapons, select the weapons, and approve the targets?
- Are there types of targets that we believe should not be attacked using cyber weapons? Do we attack them anyway if similar U.S. facilities are hit first by cyber or other weapons?
- How do we signal our intentions with regard to cyber weapons in peacetime and in crisis? Are there ways that we can use our possession of cyber weapons to deter an opponent?
- If an opponent is successful in launching a widespread, disabling attack on our military or on our economic infrastructure, how does that affect our other military and political strategies?

Didn't do too well finding the answers anywhere in U.S. government documents, congressional hearings, or officials' speeches? I didn't, either. To be fair, these are not easy questions to answer, which is, no doubt, part of the reason they have not yet been knitted together into a strategy. As with much else, how one answers these and other questions will depend upon one's experience and responsibilities, as well as the perspective that both create. Any general would like to be able to flip a switch and turn off the opposing force, especially if the same cannot be done to his forces in return. Modern generals know, however, that militaries are one of many instruments of the state, and the ultimate success of a military is now judged not just by what it does to the opponent, but by how well it protects and supports the rest of the state, including its underpinning economy. Military leaders and diplomats have also learned from past experiences that there is a fine line between prudent preparation to defend oneself and provocative activities that may actually increase the probability of conflict. Thus, crafting a cyber war strategy is not as obvious as simply embracing our newly discovered weapons, as the U.S. military did with nuclear weapons following Hiroshima.

It took a decade and a half after nuclear weapons were first used before a complex strategy for employing them, and, better yet, for not using them, was articulated and implemented. During those first years of the nuclear weapons era, accidental war almost occurred several times. The nuclear weapons strategy that eventually emerged reduced that risk significantly. Nuclear war strategy will be referenced a lot in this and the next chapter. The big differences between cyber war and nuclear war are obvious, but some of the concepts developed in the creation of nuclear war strategy have applicability to this new field. Others do not. Nonetheless, we can learn something about how a complex strategy for using new weapons can be developed by reviewing what went on in the 1950s and 1960s. And, where appropriate, we can borrow and adapt some of those concepts as we try to piece together a cyber war strategy.

THE ROLE OF DEFENSE IN OUR CYBER WAR STRATEGY

I asked at the beginning of this book: Are we better off in a world with cyber weapons and cyber war than in a theoretical world in which they never existed? The discussion in the ensuing chapters demonstrated, at least to me, that as things stand today the United States has gaping new vulnerabilities because others have cyber war capabilities. Indeed, because of its greater dependence on cyber-controlled systems and its inability thus far to create national cyber defenses, the United States is currently far more vulnerable to cyber war than Russia or China. The U.S. is more at risk from cyber war than are minor states like North Korea. We may even be at risk some day from nations or nonstate actors lacking cyber war capabilities, but who can hire teams of highly capable hackers.

Put aside for the moment the question of how it would start and consider a U.S.-Chinese cyber war as an example. We might have better offensive cyber weapons than others, but the fact that we might be able to turn off the Chinese air defense system will give most Americans limited comfort if in some future crisis the cyber warriors of the People's Liberation Army have kept power off in most American cities for weeks, shut the financial markets by corrupting their data, and created food and parts shortages nationwide by scrambling the routing systems at major U.S. railroads. Although much of China is highly advanced, a lot of it is still far from dependent upon networks controlled in cyberspace. The Chinese government may also have to worry less about temporary inconveniences experienced by its citizens or the political acceptability of measures it might impose in an emergency.

Net/net, cyber war puts America at a disadvantage right now. Whatever we can do to "them," chances are they can do more to us. We need to change that situation.

Unless we reduce our vulnerabilities to cyber attack, we will suffer from self-deterrence. Our knowing about what others could do to us may create a situation in which we are reluctant to use our superiority in other areas, like conventional weapons, in situations where it might be warranted for us to get involved. Other nations' cyber weapons may deter us from acting, not just in cyberspace but in other ways as well. In future scenarios, like ones involving China and Taiwan, or China and the offshore oil dispute, will an American President really still have the option of sending carrier battle groups to prevent Chinese action? What President would order the Navy into the Taiwan Straits, as Clinton did in 1996, if he or she thought that a power blackout that had just hit Chicago was a signal and that blackouts could spread to every major American city if we got involved? Or maybe the data difficulties the Chicago Mercantile Exchange might have just experienced could happen to every major financial institution? Worse yet, what if the Chairman of the Joint Chiefs tells the President that he does not really know whether the

Chinese can launch a damaging cyber attack that would leave the carrier battle group sitting helpless in the water? Would the President run the risk of deploying our naval superiority if trying to do so might only demonstrate that an opponent can shut down, blind, or confuse our forces?

The fact that our vital systems are so vulnerable to cyber war also increases crisis instability. As long as our economic and military systems are so obviously vulnerable to cyber war, they will tempt opponents to attack in a period of tensions. Opponents may think that they have an opportunity to reshape the political, economic, and military balance by demonstrating to the world what they can do to America. They may believe that the threat of even greater damage will appear credible and will prevent a U.S. response. Once they do launch a cyber attack, however, the U.S. leadership may feel compelled to respond. That response might not be limited to cyberspace, and the conflict could quickly escalate and get out of control.

These current circumstances argue for rapidly taking steps to reduce the strategic imbalance in which the U.S. is disadvantaged by the advent of cyber war capabilities. The answer is not to just add to our cyber offensive superiority. More U.S. cyber attack capability is unlikely to improve the imbalance or end the potential crisis instability. Unlike in conventional war, a superior offense cannot be certain to find and destroy all of the opponent's offensive capability. The tools needed to cripple the U.S. may already be in the U.S. They may not even have entered America through cyberspace, where they might be discovered, but rather on CDs in diplomatic pouches, or in USB thumb drives in businessmen's briefcases.

What is needed to reduce the risk that a nation-state will threaten to use cyber weapons against us in a crisis is for the U.S. to have a credible defense. We must cast so much doubt in the mind of the potential attacker that an attack will work against our defenses that they are he would be deterred from trying it. We want potential opponents to think that their cyber arrows might just bounce off our shields. Or at least they should think that enough of our key systems are sufficiently protected that the damage they can do to us will not be decisive. We are a long way from there today.

Defending the U.S. from cyber attacks should be the first goal of a cyber war strategy. After all, the primary purpose of any U.S. national security strategy is the defense of the United States. We do not develop weapons for the purpose of extending our hegemony over various domains (the seas, outer space, cyberspace), but as a way to safeguard the nation. While that seems simple enough, it gets complicated quickly because there are those who believe that the best way in which to defend is to attack and destroy the opponent before they can inflict damage on us.

When General Robert Elder was commander of the Air Force Cyberspace Command he told reporters that although his command has a defensive responsibility, it planned to disable an opponent's computer networks. "We want to go in and knock them out in the first round," he said. This is reminiscent of another Air Force general, Curtis LeMay, who in the 1950s, as commander of Strategic Air Command, explained to RAND Corporation analysts that his bombers would not be destroyed on the ground by a Soviet attack because "we're going first."

That kind of thinking is dangerous. If we do not have a credible defense strategy, we will be forced to escalate in a cyber conflict very quickly. We will need to be more aggressive in getting our adversary's systems so that we can stop their attacks before they reach our undefended systems. That will be destabilizing, forcing us to treat potential adversaries as current ones. We will also need to take a stronger declaratory posture to try to deter attacks on our systems by threatening to "go kinetic" in response to a cyber attack, and it will be more likely that our adversaries will think they can call that bluff.

One reason that many U.S. cyber warriors think that the best defense is a good offense is their perception of how difficult it would be to defend only by protecting. The military sees how extensive the important targets are in America's cyberspace and throws up its hands at the task of defending them all. Besides, they note (conveniently) that the U.S. military does not have the legal authority to defend privately owned and operated targets in the United States such as banks, power companies, railroads, and airlines.

This argument is the same one the Bush Administration made about Homeland Security after 9/11: that it would be too expensive to defend the U.S. against terrorists at home, so we needed to go to "the source." That thinking has had us knee deep in two wars for the last decade at a cost projected to reach $2.4 trillion, and has already cost over 5,000 American lives.

It's axiomatic that there is no single measure (or, as many in the Pentagon like to say, in a nod to the cowboy known as the Lone Ranger, no "silver bullet") that could secure U.S. cyberspace. There may, however, be a handful of steps that would protect enough of the key assets, or at least throw doubt into the mind of a potential attacker, by making it very difficult to stage a successful large-scale cyber assault on America.

Protecting every computer in the U.S. from cyber attack is hopeless, but it may be possible to sufficiently harden the important networks that a nation-state attacker would target. We need to harden them enough that no attack could disable our military's ability to respond or severely undermine our economy. Even if our defense is not perfect, these hardened networks may be able to survive sufficiently, or bounce back quickly enough, so that the damage done by an attack would not be crippling. If we can't defend every major system, what do we protect? There are three key components to U.S. cyberspace that must be defended, or, to borrow another phrase from nuclear strategy, a "triad."

THE DEFENSIVE TRIAD

Our Defensive Triad strategy would be a departure from what Clinton, Bush, and now Obama have done. Clinton in his National Plan and Bush in his National Strategy both sought to have every critical infrastructure defend itself from cyber attack. There were eventually eighteen industries identified as critical infrastructures, ranging from electric power and banking to food and retail. As previously noted, all three Presidents "eschewed regulation" as a means of reducing cyber vulnerabilities. Little happened. Bush, in the last of his eight years in office, approved an approach to cyber war that largely ignored the privately owned and operated infrastructures. It focused on defending government systems and on creating a military Cyber Command. Obama is implementing the Bush plan, including the military command, with little or no modification to date.

The Defensive Triad Strategy would use federal regulation as a major tool to create cyber security requirements, and it would, at least initially, focus defensive efforts on only three sectors.

First is the backbone. As noted in chapter 3, there are hundreds of Internet service provider companies, but only a half dozen or so large ISPs provide what is called the backbone of the Internet. They include AT&T, Verizon, Level 3, Qwest, and Sprint. These are the "trunks," or Tier 1 ISPs, meaning that they can connect directly to most other ISPs in the country. These are the companies that own the "big pipes," thousands of miles of fiber-optic cable running across the country, into every corner of the nation, and hooking up with undersea fiber-optic cables to connect to the world. Over 90 percent of Internet traffic in the U.S. moves on these Tier 1's, and it is usually impossible to get to anyplace in the U.S. without traversing one of these backbone providers. So, if you protect the Tier 1's, you are worrying about most of the Internet infrastructure in the U.S. and also other parts of cyberspace.

To attack most private-sector and government networks, you generally have to connect to them over the Internet and specifically, at some point, over the backbone. If you could catch the attack entering the backbone, you could stop it before it got to the network it was going to attack. If you did that, you would not have to worry as much about hardening tens of thousands of potential targets for cyber attack. Think about it this way: if you knew someone from New Jersey was going to drive a truck bomb into a building in Manhattan, you could defend every important building on the island (have fun getting agreement on which ones those would be), or you could inspect all trucks before they went on one of the fourteen bridges or into the four tunnels that connect to the island.

Inspecting all the Internet traffic about to enter the backbone theoretically poses two significant problems, one technical and one of policy. The technical problem is, simply, this: there is a lot of traffic and no one wants you slowing it down to look for malware or attack scripts. The policy problem is that no one wants you reading their e-mails or webpage requests.

The technical issue can be overcome with existing technology. As speeds increase, there could be difficulty scanning without introducing delay if the scanning technology failed to keep pace. Today, however, several companies have demonstrated hardware/software combinations that can scan what moves on the Internet, the small packets of ones and zeros that combine to make an e-mail or webpage. The scanning can be done so fast that it introduces no measurable delay in the packets' speeding down the fiber-optic line. And it is not just the "to" and "from" lines on the packets, the so-called headers, that would be examined, but the data level, where the malware would be. This capability is described as "deep-packet inspection," and the speed is called "line rate." The absence of delay is called "no latency." We can now do deep-packet inspection at a line rate with no latency. So the technical hurdle has been met, at least for now.

The policy problem can also be solved. We do not want the government or even an ISP reading our e-mails. The system of deep-packet inspection proposed here would be fully automated. It would not be looking for keywords, but only at the payload to see if there are predetermined patterns of ones and zeros that match up with known attack software. It's looking for signatures. If it finds an attack, it could just "black hole" the packets, dump them into cyber oblivion, or it could quarantine them, put them aside for analysis. For Americans to be satisfied that such a deep-packet inspection system were not Big Brother spying on us, it would have to be run by the Tier 1 ISPs themselves and not by the government. Moreover, there would have to be rigorous oversight by an active Privacy and Civil Liberties Protection Board to ensure that neither the ISPs nor the government was illegally spying on us.

The idea of putting deep-packet inspection systems on the backbone does not create the risk of government spying on us. That risk already exists. As we saw with the illegal wiretapping in the Bush Administration, if the checks and balances in the system fail, the government can already improperly monitor citizens. That is a major concern and needs to be prevented by real oversight mechanisms and tough punishment for those who break the law. Our nation's strong belief in privacy rights and civil liberties is not incompatible with what we need to do to defend our cyberspace. Giving guns to police does raise the possibility that some policemen may get involved in unjust shootings on rare occasions, but we recognize that we need armed police to defend us and we work hard at making sure that unjust shootings are prevented. So, too, we can deploy deep-packet inspection systems on Internet backbone ISPs, recognizing that we need them there to protect us, and we have to make sure that they do not get misused.

How would such a system get deployed? The deep-packet inspection systems would be placed where fiber-optic cables come up out of the ocean and enter the U.S., at "peering points," where the Tier 1 ISPs connect to each other and the smaller networks, and at various other points on the Tier 1 networks. The government, perhaps Homeland Security, would probably have to pay for the systems, even though they would be run by the ISPs and maybe systems integrator companies. The signatures of the malware that the black box scanners would look for would come from Internet security companies such as Symantec and McAfee, which have elaborate global systems to look for malware. The ISPs and government agencies could also provide signatures.

The black box inspectors would have to be connected to each other on a closed network, what is called "out-of-band communications" (not on the Internet), so that they could be updated quickly and reliably even if the Internet were experiencing difficulties. Imagine that a new piece of attack software enters into cyberspace, one that no one has ever seen before. This "Zero Day" malware begins to cause a problem by attacking some sites. The deep-packet inspection system would be tied into Internet security companies, research centers, and government agencies that are looking for Zero Day attacks. Within minutes of the malware being seen, its signature would be flashed out to the scanners, which would start blocking it and would contain the attack.

A precursor to this kind of deep-packet inspection system is already being deployed. Verizon and AT&T can, at some locations, scan for signatures that they have identified, but they have been reluctant to "black hole" (or kill) malicious traffic because of the risk that they might be sued by customers whose service is interrupted. The carriers would probably win any such suit because their service-level agreements (SLAs) with their customers usually state that they have the right to deny service if the customer's activity is illegal or disruptive to the network. Nonetheless, because of the typical abundance of caution from their lawyers, the companies are doing less than they could to secure cyberspace. Legislation or regulation is probably needed to clarify the issue.

The Department of Homeland Security's "Einstein" system, discussed in chapter 4, has been installed at some of the locations where government departments connect to the Tier 1 ISPs. Einstein only monitors government networks. The Defense Department has a similar system at the sixteen locations where the unclassified DoD intranet connects to the public Internet.

A more advanced system, with higher speed capacity, more memory and processing capabilities, and out-of-band connectivity, could help to minimize or deter a large-scale cyber attack if it were broadly deployed to protect not just the government, but the backbone on which all networks rely. By defending the backbone in this way, we should be able to stop most attacks against our key government and private-sector systems. The independent Federal Communications Commission has the authority today to issue regulations requiring the Tier 1 ISPs to establish such a protective system. The Tier 1's could pass along the costs to their customers and to smaller ISPs that peer with them. Alternatively, Congress could appropriate funds for some or all of the system. So far, the government is only beginning to move in this direction, and then only to protect itself, not the private-sector networks on which our economy, government, and national security rely.

ISPs should also be required to do more to keep our nation's portion of the cyber ecosystem clean. Ed Amoroso, the chief security officer at AT&T, told me that his security operations center watches as computers that have been taken over by a botnet spew out DDOS and spam. They know what subscribers are infected, but they don't dare inform the customer (much less cut off access) out of fear that customers would switch providers and try to sue them for violating their privacy. That equation needs to be stood on its head. ISPs should be *required* to inform customers of the network when data shows that their computers have been made part of a botnet. ISPs should be required to shut off access if customers do not respond after being notified. They should also be required to provide free antivirus software to their subscribers, as many now do because it helps them manage their bandwidth better; and subscribers should be required to use it (or whatever antivirus software they choose). We don't let car manufacturers sell cars without seat belts, and in most states we don't let people drive cars unless they are wearing them. The same logic should apply on the Internet, because poor computer security by an individual creates a national security problem for us all.

In addition to the Tier 1 carriers screening Internet traffic at packet level for known malware, blocking those packets that match previously identified attacks, related steps could be added to strengthen the system. First, with relatively little investment of money and time, software could be developed to identify "morphed malware." The software would look for slight variations in known attack signatures, changes that attackers might use in attempts to slip by the deep packet inspection of previously identified hacks. Second, in addition to having the Tier 1 ISPs looking for malware, government and large, regulated commercial institutions such as banks would also contract with hosting and data centers to do deep-packet inspection. At a handful of large hosting data centers scattered around the country, the fibers of Tier 1 ISPs come together to do switching among the networks. At these locations, some large institutions also have their own servers locked behind fencing in row upon row of blinking equipment or stashed in highly secured rooms. The operators of these centers can screen for known malware as a second level of defense. Moreover, the data center operators or IT security firms can also look at data after it has passed by. The data centers can provide managed security services, looking for anomalous activity that might be caused by previously unidentified malware. Unlike the attempts to block known malware as it comes in, the managed security services would look for patterns of suspicious behavior and anomalous activity of data packets over time. By doing that, they may be able to spot more complicated two-step attacks and new Zero Day malware. That new malware would then be added to the list of things to be blocked. Searches could be performed for locations in the data banks where the new malware had gone, perhaps allowing the system to stop large-scale exfiltration of data.

By paying the ISPs and managed security service providers to do this sort of data screening, the government would remain sufficiently removed from the process to protect privacy and to encourage competition. The government's role, in addition to paying for the defenses, would be to provide its own information about malware (locked up in a black box if necessary), incentivize firms to discover attacks, and create a mechanism to allow the public to confirm that privacy information and civil liberties are well protected. Unlike a single line of defense owned and operated by the government (such as the "Einstein" system being created by the Department of Homeland Security to protect civilian federal agencies), this would be a multilayer, multiple-provider system that would encourage innovation and competition among private sector IT companies. If the government was aware that a cyber war was about to break out, or if one already had, a series of federal network operation centers could interact with these private IT defenders and with the network operation centers (NOCs) of key privately owned institutions to coordinate a defense. For that to happen, the government would have to create in advance a dedicated communications network among the NOCs, one that was highly secure, entirely separated, and in other ways different from the Internet. (The fact that such a new network would be needed should tell you something about the Internet.)

The second prong of a Defensive Triad is a secure power grid. The simplest way to think about this idea is to ask, as some have, why the hell is the power grid connected to cyberspace at all, anyway? Without electricity, most other things we rely on do not work, or at least not for long. The easiest thing a nation-state cyber attacker could do today to have a major impact on the U.S. would be to shut down sections of the Eastern or Western Interconnects, the two big grids that cover the U.S. and Canada. (Texas has its own, third, grid). Backup power systems are limited in duration and notorious for not coming on when needed (as happened at my house last night when a lightning storm hit the rural power net, creating a localized black-out. My automatic starting generator sat there like an oversized door stop). Could those three North American power-sharing systems, composed of hundreds of generation and transmission companies, be secured?

Yes, but not without additional federal regulation. That regulation would be focused on disconnecting the control network for the power generation and distribution companies from the Internet and then making access to those networks require authentication. It would really not be all that expensive, but try telling that to the power companies. When asked what assets of theirs were critical and should be covered by cyber security regulations, the industry replied that 95 percent of their assets should be left unregulated with regard to cyber security. One cyber security expert who works with the major cyber security auditing firms said he asked each audit firm that had worked with power companies if they had been able in their audits to get to the power grid controls from the Internet. All six firms said they had. How long did it take them? None had taken longer than an hour. That hour was spent hacking into the company's public website, then from there into the company's intranet, then through "the bridge" they all have to their control systems. Some audits cut the time by hacking into the Internet-based phones (voice over Internet protocol, or VOIP, phones) that were sitting in the control rooms. These phones are by definition connected to the Internet; that's how they connect to the telephone network. If they are in the control room, they are also probably connected to the network that runs the power system. Good thinking, huh? Oh, it gets better. In some places the commands to electrical grid components are sent in the clear (that is, unencrypted) via radio, including microwave. Just sit nearby, transmit on the same frequency with more energy in your signal than the power company is using, and you are giving the commands (if you know what the command software looks like).

The Federal Energy Regulatory Commission (FERC) promises that in 2010 it really will start penalizing power companies that do not have secure cyber systems. What they have not said is how the Commission will know who is in violation, since the FERC doesn't have the staff to regularly inspect. The U.S. Department of Energy, however, has hired two cyber security experts to determine if the $3.4 billion in Smart Grid grants are going to new programs that are adequately secured. Smart Grid is the Obama Administration's idea to make the power grid even more integrated and digitized. Power companies can ask for some of that money by submitting proposals to the Energy Department. When they do, the two experts will read the proposals to see if there is a section somewhere that says "cyber security." The Energy Department refuses to say who the two experts are or what they will be looking for in the "cyber security" section of the grant proposal. There are no publicly available standards. One idea for a standard might be that the taxpayers don't give any of the $3.4 billion in Smart Grid money to companies that haven't secured their current systems. Don't expect the Energy Department to use that standard anytime soon, because that would mean taking advantage of this unique federal giveaway program to incentivize people to make things more secure. That smacks of regulation, which, of course, is just like socialism, which is un-American. So, we will soon have a more digital Smart Grid, which will also be a Less Secure Grid. How could we make the U.S. national electrical system a Smart *and* Secure Grid?

The first step in that direction would be issuing and enforcing serious regulations to require electric companies to make it next to impossible to obtain unauthorized access to the control network for the power grid. That would mean no pathway at all from the Internet to the control system. In addition, the same kind of deep-packet inspection boxes I proposed placing on the Internet backbone could be placed on the points where the control systems link to the power companies' intranets. Then, just to make things even harder for an attacking cyber warrior, we could require that the actual control signals sent to generators, transformers, and other key components be both encrypted and authenticated. Encrypting the signals would mean that even if you could hack your way in and try to give an instruction to a generator, you would not have the secret code to do so. Authenticating the commands would mean that through a proof of identity procedure, or electronic "handshake," the generator or transformer would know for sure that the command signal it was getting was coming from the right place. Because some parts of the grid might still be taken over by a nation-state hacker, certain key sections should have a backup communication system for sending command and control signals so that they could restore service.

Many people dismiss the significance of an attack on the power grid. As one senior U.S. government official said to me, "Power blackouts take place all the time. After a few hours, the lights come back." Maybe not. The power comes back after a few hours when what has caused it to fail is a lightning storm. If the failure is the result of intentional activity, it will likely be a much longer blackout. In what is known as the "Repeated Smackdown Scenario," cyber attacks take down the power grid, and keep it down for months.

If the attacks destroy generators, as in the Aurora tests, replacing them can take up to six months, because each must be custom built. Having an attack take place in many locations simultaneously, and then happen again when the grid comes back up, could cripple the economy by halting the distribution of food and other consumer goods, shutting down factories, and forcing the closure of financial markets.

Do we really need improved regulation? Should we force power companies to spend more to secure their networks? Is the need real? Let's ask the head of U.S. Cyber Command, General Keith Alexander, the man whose cyber warriors would attack other nations' electric grids. Knowing what he knows he can do to others, does the General think we need to do more to protect our own power grid? That's essentially what he was asked in a congressional hearing in 2009. He replied, "So the power companies are going to have to go out and change the configuration of their networks. . . . [T]o upgrade their networks to make sure they are secure is a jump in cost for them. . . . And now you're going to have to work through their regulatory committees to get the rate increases so that they can actually secure their networks. . . . [H]ow does government, because we're interested in perhaps having reliable power, how do we ensure that that happens as a critical infrastructure?" It was a little rambling, but General Alexander seemed to be saying that power companies need to reconfigure so we can have secure, reliable electricity, that this may mean they have to spend more, and that the regulatory organizations will have to help make that happen. He's right.

The third prong of the Defensive Triad is Defense itself, as in the Department of Defense. There is little chance that a nation-state would stage a major cyber attack against the U.S. without trying to cripple DoD in the process. Why? While a nation-state actor might try to cripple our country and our will by destroying private-sector systems like the power grid, pipelines, transportation, or banking, it is hard to imagine such actions coming as a bolt from the blue. Cyber attacks would only likely come in a period of heightened tensions between the U.S. and the attacker nation. In such an atmosphere, the attacker would probably already fear the possibility of conventional, or kinetic, action by the U.S. military. Moreover, if an opponent were going to hit us with a large cyber attack, they would have to assume that we might respond kinetically. A cyber attack on the U.S. military would likely concentrate on DoD's networks.

For simplicity, let's say that there are basically three DoD networks. The first, NIPRNET, is the unclassified intranet. Systems on that network use the dot-mil addresses. The NIPRNET connects to the public Internet at sixteen nodes. While it is unclassified data that moves on NIPRNET, unclassified does not mean unimportant. Most logistical information, like supplying Army units with food, is on the NIPRNET. Most U.S. military units cannot sustain themselves for long without support from private-sector companies, and most of that communication goes through the NIPRNET.

The second DoD network is called SIPRNET and is used to pass secret-level classified information. Many military orders are transmitted over the SIPRNET. There is supposed to be an "air gap" between the unclassified and secret-level networks. Users of the classified network download things from the Internet and upload them to the SIPRNET, thus sometimes passing malware along unknowingly. Pentagon information security specialists call this problem the "sneakernet threat."

In November 2008, a Russian-origin piece of spyware began looking around cyberspace for dot-mil addresses, the unclassified NIPRNET. Once the spyware hacked its way into NIPRNET computers, it began looking for thumb drives and downloaded itself onto them. Then the "sneakernet effect" kicked in. Some of those thumb drives were then inserted by their users into classified computers on the SIPRNET. So much for the air gap. Because the secret network is not supposed to be connected to the Internet, it is not supposed to get viruses or worms. Therefore, most of the computers on the network had no antivirus protection, no desktop firewalls or similar security software. In short, computers on DoD's most important network had less protection than you probably have on your home computer.

Within hours, the spyware had infected thousands of secret-level U.S. military computers in Afghanistan, Iraq, Qatar, and elsewhere in the Central Command. Within a few more hours, the highest-ranking U.S. military officer, Admiral Mike Mullen, the Chairman of the Joint Chiefs of Staff, was realizing how vulnerable his military really was. According to a high-ranking Pentagon source, Mullen screamed, "You mean to tell me that I can't rely on our operational network?" at the network specialists briefing him. The network experts on the Joint Staff acknowledged the Admiral's conclusion. They did not seem surprised; hadn't he known that already? Horrified at a huge weakness that Majors and Captains seemed to take for granted, but which had been kept from him, Mullen looked around for a senior officer. "Where's the J-3?" he demanded, looking for the Director of Operations. "Does he know this?"

Shortly thereafter, Mullen and his boss, Secretary of Defense Robert Gates, were explaining their discovery to President Bush. The SIPRNET was probably compromised. The netcentric advantage the U.S. military thought it enjoyed might just prove to be its Achilles' heel. Perhaps Mullen should not have been surprised. There are over 100,000 SIPRNET terminals around the world. If you can get time alone with one terminal for a few minutes, you can upload malware or run a covert connection to the Internet. One friend of mine described a SIPRNET terminal in the Balkans that a Russian "peacekeeper" could easily get to without being observed.

Just as in World War II, when the Allies needed only one German Enigma code machine in order to break the Nazis' encryption, so, too, if one SIPRNET terminal is compromised, can malware be inserted that could affect the entire network. Several experts who worked on SIPRNET security-related issues confirmed to me the scary conclusion. As one said, "You got to assume that it's not going to work when we need it." He explained that if, in a crisis, that command and control network were brought down by an enemy, or, worse, if the enemy issued bogus commands, "the U.S. military would be severely disadvantaged." That's putting it mildly.

The third major DoD network is the Top Secret/Sensitive Compartmented Information (TS/SCI) network called JWICS. This more limited network is designed to pass along intelligence information to the military. Its terminals are in special highly secured rooms known as Secret Compartmentalized Information Facilities, or SCIFs. People also refer to those rooms as "the vault." Access to these terminals is more restricted because of their location, but the information flowing on the network still has to go across fiber-optic cables and through routers and servers, just as with any other network. Routers can be attacked to cut communications. The hardware used in computers, servers, routers, and switches can all be compromised at the point of manufacture or later on. Therefore, we cannot assume that even this network is reliable.

Under the CNCI plan, DoD is embarked on an extensive program to upgrade security on all three kinds of networks. Some of what is being done is classified, much of it is expensive, and some of it will take a long time. A real possibility is the use of high-bandwidth lasers to carry communications to and from satellites. Assuming the satellites were secure from hacking, such a system would reduce the vulnerabilities associated with fiber-optic cable and routers strung out around the world. There are, however, a few important design concepts using currently available technology that should be included in the DoD upgrade program quickly, and they are not budget busters:

- in addition to protecting the network itself, guard the end points; install desktop firewalls and antivirus and intrusion-prevention software on all computers on all DoD networks, whether or not they are connected to the Internet;
- require all users on all DoD networks to prove who they are when they sign on through at least two factors of authentication;
- segment the networks into subnets with limited "need to know" access rules for connecting out of the subnets;
- go beyond the current limited practice of bulk encrypting, which scrambles all traffic as it moves on trunk fiber cables, and encrypt all files on all computers, including data at rest in data-storage servers;
- monitor all networks for new unauthorized connections to the network, automatically shutting off unknown devices.

Even if its networks are secure, DoD runs the risk that the software and/or hardware it has running its weapons systems may be compromised. We know the plans for the new F-35 fighter were stolen by hack into a defense contractor. What if the hacker also added to the plans, perhaps a hidden program that causes the aircraft to malfunction in the air when it receives a certain command that could be radioed in from an enemy fighter? Logic bombs like that can be hidden in the millions of lines of code on the F-35, or in the many pieces of firmware and computer hardware that run the aircraft. As one pilot told me, "Aircraft these days, whether it's the F-22 Raptor or the Boeing 787 . . . all they are is a bunch of software that happens to be flying through the air. Mess with the software and it stops flying through the air." I thought of the Air France Airbus that crashed in the South Atlantic because its computer made a wrong decision.

The computer chips U.S. weapons use, as well as some of the computers or their components, are made in other countries. DoD's most ubiquitous operating system is Microsoft Windows, which is developed around the world on development networks that have proven vulnerable in the past. This supply-chain concern is not easily or quickly solved. It is one of the areas that the 2008 Bush plan focused on. New chip factories, or fabs, are being built in the U.S. Some private-sector companies are developing software to check other software for bugs. In addition to adding quickly to the security of its networks, one of the most important things the Pentagon could do would be to develop a rigorous standards, inspection, and research program to ensure that the software and hardware being used in key weapons systems, in command control, and in logistics are not laced with trapdoors or logic bombs.

So that's the Defensive Triad strategy. If the Obama Administration and the Congress were to agree to harden the Internet backbone, separate and secure the controls for the power grid, and vigorously pursue security upgrades for Defense IT systems, we could cast doubt in the minds of potential nation-state attackers about how well they would do in launching a large-scale attack against us. And even if they did attack, the Defensive Triad could mitigate the effects. It is admittedly difficult to measure the financial cost of these programs at this point in their development, but in terms of implementation difficulty, they could all be phased in over five years. If implemented with the thought in mind that we want to be able to derive some benefit from the improvements even before they are fully deployed, there could be a steady increase over those five years in the degree of difficulty for a nation-state thinking about cyber war against us. Unless and until this plan or some similar defensive strategy that includes the private-sector networks is implemented, being in a cyber war would probably not be good news for the United States.

If we do the Defensive Triad, we will have the credibility to say some things that will add further to our ability to deter cyber attack. Sometimes just saying things, things that do not always cost money, can buy you added security, if you have credibility. The capstone of the triad is our "declaratory posture" toward those nation-states that would think about attacking us through cyberspace. A declaratory posture is a formally articulated statement of the policy and intention of the government. We do not have an authoritatively articulated policy today about how we would regard a cyber attack and what we would do in response. Some in the councils of a potential attacker could argue that the U.S. response to a cyber attack might be fairly minimal, or confused.

We do not want to be in a situation similar to what John Kennedy found himself in *after* he discovered that there were nuclear-armed missiles in Cuba. He declared that any such missile fired by anyone (Russian or Cuban) from Cuba toward "any nation in this hemisphere would be regarded as an attack, by the Soviet Union, upon the United States, requiring a full retaliatory response." Those words were chilling when I first heard them as a twelve-year-old; they remain so today. If the U.S. had said that before the missiles went to Cuba, the Kremlin might not have sent them.

A public declaration about what we would do in case of a cyber attack should, however, not limit future decisions. There needs to be a certain "constructive ambiguity" in what is said. In the event of a major cyber attack, there will likely be an unhelpful ambiguity about who attacked us, and our declaratory policy needs to take that into account as well. Imagine, then, Barack Obama addressing the graduating class of one of the four U.S. military academies, something he will do four times in his first term in office. He looks out on the sea of uniformed new officers and their parents, describes the phenomenon of cyber war, and then says: "So let me make this clear to any nation that may contemplate using cyber weapons against us. The United States will regard a cyber attack that disrupts or damages our military, our government, or our critical infrastructure as we would a kinetic attack that had the same target and the same effect. We would consider it a hostile act in our territory. In response to such aggression in our cyberspace, I, as Commander in Chief, will draw upon the full panoply of power available to the United States of America and will not be limited as to the size or nature of our response by those characteristics of the attack upon us."

"Panoply of power" is a presidential phrase. It says he may respond with diplomatic, economic, cybernetic, or kinetic means, as he chooses and as appropriate, taking into account the target and the effect. International lawyers will quibble about the "not be limited" line, noting that defensive responses are supposed by international law conventions to be commensurate with the attack. Suggesting the response might be incommensurate, however, adds to deterrence. In nuclear strategy this idea was called "escalation dominance"—responding to a lower-level attack by moving rapidly up the escalation ladder and then saying that the hostilities must end. It sends the message that you are not willing to engage in some prolonged, slow-bleeding conflict. It is an option that the President must have, whether or not he uses it.

What if, as is likely, the attribution problem occurs and the attacker hides behind the skirts of "citizen hacktivists" or claims the attack merely transited their country, but did not originate there? Anticipating this claim in advance, Obama pauses in his address and then adds, "Nor will we be fooled by claims that a cyber attack was the work of citizen hacktivists or that attribution is uncertain. We have the capability to determine attribution to the degree necessary. Moreover, we reserve the right to consider a refusal to stop, in a timely manner, an attack emanating from a country as the equivalent of the government of that country engaging in the attack. We will also judge a lack of serious cooperation in investigations of attacks as the equivalent of participation in the attack."

The Obama Doctrine would be one of *cyber equivalency*, in which cyber attacks are to be judged by their effects, not their means. They would be judged as if they were kinetic attacks, and may be responded to by kinetic attacks, or other means. The corollary is that nations have a *national cyberspace accountability* and an *obligation to assist*, meaning that they would have a responsibility to prevent hostile action coming from servers in their country and must promptly hunt down, shut off, and bring to justice those who use their cyberspace to disrupt or damage systems elsewhere. America would also have these obligations and would have to shut off botnets attacking nations like Georgia from places like Brooklyn. If the Tier 1 ISPs were scanning their networks, the *obligation to assist* would be fairly easy to carry out.

Were Obama or a future President to articulate such a doctrine, the United States would have made clear that it regarded cyber attacks that disrupt or damage things not as a lesser, more permissible form of national action just because they may not result in colorful explosions or in piles of body bags. If the President also adopted something like the Defensive Triad, the U.S. would finally have a credible cyber war defensive strategy.

So, once we have reasonable defenses in place, would we then be able to go on the offensive, using our new cyber warriors to achieve military dominance of cyberspace for the United States of America?

CHAPTER SIX

HOW OFFENSIVE?

In the seminal 1983 movie about computers and war, *War Games*, starring a young Matthew Broderick, the tinny computer voice asked haltingly, "Do you want to play a game of thermonuclear war?" Why don't we play a game of cyber war in order to elucidate some of the policy choices that shape a strategy. DoD runs such exercises, called Cyber Storm, annually. The CIA's annual cyber war exercise, Silent Horizon, has been happening since 2007. For the purposes of this analysis, I'll make the same request of you that I made of students at Harvard's Kennedy School and national security bureaucrats sitting around the White House Situation Room conference table: "Don't fight the scenario." By that I mean, do not spend a lot of time rejecting the premise that circumstances could happen someday that would result in the U.S. being on the edge of conflict with Russia or China.

When U.S. cyber warriors talk about the "big one," they usually have in mind a conflict in cyberspace with Russia or China, the two nations with the most sophisticated offensive capability other than the U.S. No one wants hostilities with those countries to happen. Thinking about it, for the purposes of understanding what cyber war would look like, does not make it more likely. In fact, by understanding the risks of our current cyber war posture, we might reduce the chances of a real cyber war. And if, despite our intentions, a cyber war does happen, it would be best to have thought in advance about how it could unravel.

Certainly, I did not want to see the attack of 9/11 happen, but I had chaired countless "tabletop exercises," or war game scenarios, to get myself and the bureaucracy ready in case something like it did happen. When it came, we had already thought through how to respond on the day of an attack and the few days thereafter. We spent enormous effort to try to prevent attacks, but we also devoted some time to thinking about what we would do if one succeeded. Had we not done so, that awful day would have been even worse. So, in that spirit of learning by visualizing, let's think about a period of rising tensions between the U.S. and China.

Let's call it Exercise South China Sea and set it a few years in the future. Not much has changed, except China has increased its dependence on the Net somewhat. For its part, the U.S. has not done much to improve its cyber defenses. We will have three teams, U.S. Cyber Command, the Chinese People's Liberation Army (PLA) Cyber Division, and the Controllers, who play the part of everyone else. The Controllers also decide what happens as a result of the other two teams' moves. Let's say for the sake of the exercise that China has been aggressively pressing Vietnam and other ASEAN (Association of Southeast Asian Nations) countries to cede their rights to a vast and rich undersea area of gas and oil fields. (China has, in fact, claimed waters that run hundreds of miles to its south, along the coasts of Vietnam and the Philippines.) We will stipulate that there have been small clashes between their navies. In an irony of history, we will say that the government of Vietnam has asked the U.S. for military support, as have other nations in the region with claims on the contested waters. In response, the President has authorized a joint U.S.-ASEAN naval exercise and has dispatched two U.S. carrier battle groups, about twenty ships, including about 150 aircraft and several submarines. China and the U.S. have exchanged diplomatic notes and public pronouncements, with both countries essentially saying that the other one should stay out of the issue. American cable news networks have at this point started showing dramatic slides with the words "South China Sea Crisis."

As our hypothetical exercise opens at Fort Meade, the team playing Cyber Command has been ordered by the Pentagon to prepare a series of steps it could take as the political situation escalates. The order from the Secretary of Defense is to develop options to:

> First, dissuade the Chinese government from acting militarily over the contested waters. Second, failing that, to reduce to the maximum extent possible the ability of the Chinese military to pose a risk to U.S. and allied forces in the area. Third, in the event of increased tensions or the outbreak of hostilities, to be able to disrupt the Chinese military more broadly to reduce its ability to project force. Fourth, to occupy the Chinese leadership with disruption of their domestic infrastructure to the extent that it may cause popular and Party questioning of the Chinese government's aggressive behavior abroad. Fifth, throughout this period Cyber Command is to work with appropriate U.S. government agencies to prevent Chinese-government or Chinese-inspired cyber attacks on the U.S. military or significant U.S. infrastructure.

In this situation, the team playing Cyber Command in the table-top exercise faces a dilemma. They do not want to expose all of the cyber attack techniques, or "exploits," that they have developed. Once an exploit is used, cyber defenders will devote the time and energy necessary to figure out how to block it in the future. While the defenders will not fix all of the systems that could be exploited, they will patch enough of the important systems that the attack technique will have lost much of its potency. Thus, Cyber Command will want to withhold its most clever attacks. If they wait, however, the Chinese may have done things that make it far more difficult for the U.S. to execute cyber attacks.

As tensions begin to mount, China will reduce the flow of packets into China and will scan and filter for possible U.S. attacks the ones it permits in. Then it may drop connectivity to the outside world altogether. If the U.S. has not already launched its cyber attack, it will be much harder to get around the Great Firewall of China. Cyber Command will have to have created, in advance, tunnels into Chinese cyberspace, perhaps by hiding satellite telephones in China to download attacks and insert them into the Internet behind the Wall. Or perhaps Cyber Command will, working with CIA, have placed agents inside China with the attack tools already on their laptops.

If the U.S. waits to use its best weapons, China may make it difficult to launch an attack from U.S. cyberspace by confusing or crashing our cyberspace and Internet backbone. Scrambling data on the highest-echelon servers of the Domain Name System, which provides the Internet addresses of websites, or doing so on the routing tables (the Border Gateway Protocol lists) of the Tier 1 backbone providers will disrupt U.S. cyberspace for days. The effect would be to send traffic more or less randomly to the wrong place on the Internet. As noted in chapter 3, very little prevents this from happening now since these software programs that make the Internet run do not require that there be any checking to see if the commands issued are authentic.

If the Chinese could get agents into the big windowless buildings where all the Tier 1 ISPs link to each other, the so-called peering points, or into any place on the Tier 1 ISP networks, they could possibly issue commands directly to the routers that do the switching and directing of traffic on the Internet and in the rest of cyberspace. Even though DoD and U.S. intelligence agencies have their own channels in cyberspace separate from the public Internet, their traffic is likely to be carried on the same fiber-optic cable pipes as the public Internet. The public Internet may just be a different "color" on the same fiber or maybe a different fiber in the same pipe. Chances are that there are many places where the DoD and intelligence-agency traffic is running through the same routers as the public Internet. As discussed earlier, China is very familiar with the routers. Most of them are made by the U.S. firm Cisco, but made in China.

All of that Chinese potential to disrupt the Internet and stop the U.S. from being able to send cyber attacks out means that the Cyber Command team has an incentive in the early stages of a crisis to store their attacks on networks outside of the U.S. Of course, doing so broadens the global involvement in the pending cyber war.

To begin operations, the team playing Cyber Command decides to signal their involvement with the hope of deterring China from engaging in further military operations. The act that Cyber Command conducts must be deniable publicly, but Chinese authorities must know it was no accident. The signal must demonstrate an ability to do things that are technically hard and which are significant enough for the Chinese leadership to notice, but without being so damaging as to provoke a full-scale cyber war.

Having hacked their way into the closed Chinese military intranet, they send around to senior officers a doctored picture of China's one aircraft carrier, but in this Photoshopped version the ship is in flames and sinking. The not-so-subtle message is that the pride of China's navy, its one carrier, could easily be sunk by the 7th Fleet, causing great loss of face to the Chinese military; maybe it's better not to get into what could prove to be such an embarrassing fight.

U.S. intelligence then learns that the Chinese are loading up their South Sea Fleet for an amphibious landing on disputed islands in the South China Sea. Cyber Command is asked by the Pentagon to buy some time, to slow down the Chinese landings by disrupting the troops and supplies getting ready to load up on the ships still in port. The Chinese South Sea Fleet is headquartered in Zhanjiang, on the Leizhou Peninsula, and its air force supporting operations in the South China Sea is on Hainan, in the Tonkin Gulf. The Fleet Headquarters and the Naval Air Base do not have their own electric grid; they are connected to the public power system. They do not have their own large generators, just smaller emergency backup units.

Using its subordinate unit, the 10th Fleet, Cyber Command utilizes a preexisting trapdoor in the Chinese power grid and accesses the local electric grid's controls. Once in the control system, they issue signals that cause surges, tripping breakers that shut down transmission and stop generators. The Americans do not cause the generators or transformers to damage themselves.

The team playing China in our hypothetical exercise realizes that the blackout was caused by an intrusion and orders a trace on the attack. It is traced back to an ISP in Estonia, where the trail goes cold. No one in Beijing would think a hacker in Estonia is the real attacker. Thus the signal is sent, but in a deniable way. The signal does get the Chinese team's attention. They are informed that the blackout on the Leizhou Peninsula tripped a cascade that knocked out all of Guangdong (formerly Canton) Province, leaving slightly more than a hundred million Chinese in the dark for almost twenty-four hours. Hong Kong was also affected. The Politburo considers the blackout an escalatory step and asks the team playing the cyber warfare division of the PLA for options to respond.

The PLA team recommends China respond in a somewhat commensurate manner, going after cities with Navy bases, but they want to do more as well, to send the U.S. the message that they can hurt us more than we can hurt them. The Politburo approves all six steps proposed by their cyber warriors:

1. ordering the South Sea Fleet reinforced and moving more aircraft to Hainan and the southern coast airfields;

2. directing the Chinese submarine squadron at Yulin, on Hainan Island, to go to sea;

3. activating logic bombs already planted in the power grids in Honolulu, San Diego, and Bremerton, in the state of Washington, three cities where much of the U.S. Pacific Fleet is located (and although the Chinese do not know it, the blackouts will extend into Tijuana, Mexico, and up to Vancouver, British Columbia);

4. disrupting the unclassified DoD network by launching a new, never-before-seen worm (a Zero Day exploit) that infects one machine after another and causes their hard drives to be erased (the attack is launched from inside the DoD intranet);

5. attacking the Estonian ISP from which the attack on the Chinese power grid appears to have originated; and

6. causing a power blackout in Yokosuka and the surrounding area in Japan where the U.S. 7th Fleet is headquartered.

At the beginning of the next move in the exercise, with tensions escalating, Cyber Command is informed that China is about to stop Internet traffic from the outside world. The Fort Meade team, therefore, proposes to the Pentagon that it be authorized to launch two more waves of cyber attack and be prepared to launch a third. The two attacks would be on the Chinese air defense network and on the national military command control system. These attacks would use highly secret exploits and activate logic bombs already planted in these networks. In the wings would be a broad attack on the Chinese rail network, air traffic control, the banking system, and the hardware of the power grid (generators and transformers).

Somewhat to their surprise, the Cyber Command team receives instructions from the Control Team playing the White House and Pentagon to avoid attacks on the military command and control system and on defensive weapons like air defense. The Cyber Command planners are also told to avoid both the air traffic control system and the banking sector.

As the Cyber Command team is reformulating its next move, databases at the Security Industries Automation Corporation and the Deposit Trust in New York are reported to be seriously damaged and corrupted. Data has also been badly scrambled at CSX, Union Pacific, and the Burlington Northern Santa Fe railroads, as well as at United, Delta, and American Airlines. As a result, the New York Stock Exchange has closed, freight trains have stopped, and aircraft are sitting at gates across the country. The Defense Information Systems Agency, which runs DoD's internal networks, declares an emergency because both the secret-level SIPRNET and the top-secret JWICS networks have been disrupted by fast-spreading worms that are crashing hard drives. None of these attacks originated overseas, and therefore U.S. intelligence and Cyber Command did not see them coming and could not stop them before they got to the U.S. The attacks appear to have used new, not previously employed techniques, and thus Cyber Command was unable to block them by scanning for the signatures of past attacks.

With attacks on Chinese air defense, banking, national military command and control, and air traffic control ruled out by higher authorities, the team playing U.S. Cyber Command has fewer options than it thought it would. Moreover, because U.S. Cyber Command has a defensive role in protecting DoD networks, some of the team members are removed to deal with the worms working their destructive path through the Defense Department. In light of the significant escalation that the Chinese team utilized in its first move, the U.S. team opts to launch a nationwide power blackout in China, including targeted attacks to damage several large generators. At the same time, they will try to cause a maximum number of freight-train derailments and jumble the database of the rail system. To replace the military targets that have been ruled out by their superiors, the U.S. team decides to attack the communications satellite used by the Chinese navy and the navy's logistics network.

The Control Team's report on the effects of the second round of U.S. attacks is not good news. China had disconnected its networks from the global Internet, thus limiting the impact of the U.S. attack. Moreover, when the U.S. first attacked the power grid, Beijing had ordered all remaining sectors of the electric grid to go to a defensive posture that disconnected Internet links and broke up regional grids into "islands" to prevent cascading blackouts. Only a few of the generators targeted by the U.S. can be hit and their destruction will cause only isolated outages. At the same time that defenses were raised elsewhere, the rail system shifted to a manual, radio-based control system. Therefore the attempted second attack on the freight-rail system by the U.S. did not work.

The U.S. hacked the Chinese communications satellite, causing its station-keeping thrusters to fire until all fuel was spent and sending it in the direction of Jupiter. Within an hour, however, the Chinese navy has activated a backup, encrypted radio Teletype system. But the U.S. attack on the Chinese navy logistics computer network is successful and, together with the regional power blackout, has slowed the boarding of Chinese troops onto ships. The Control Team also reports that a Chinese submarine has surfaced between the two U.S. aircraft carriers. It had penetrated the defensive perimeter, similar to an incident that actually happened in 2009 when a Song-class submarine appeared next to the carrier USS *Kitty Hawk*. By surfacing, the sub has given away its location, but it has also sent a message to the U.S. that the location of the carriers is known with precision, making it possible for China to flood the area with air- and ground-launched cruise missiles if shooting starts.

At this point, the U.S. Cyber Command team is informed that the White House has ordered the two U.S. carrier battle groups to proceed toward Australia. The State Department will be sending a high-level team to Beijing to discuss its territorial claims. Cyber Command has been ordered to cease offensive action.

The game is over.

After every tabletop exercise in the government, there is a gathering of controllers and players called "the hot wash." It is a time to write down lessons learned and to make note of areas for further study. So what did we learn from Exercise South China Sea? What issues did it highlight? Ten important cyber war issues emerged from the players' conduct of the simulation: the use of deterrence; the concept of going first; the prewar preparation of the battlefield; the global spread of a regional conflict; collateral damage; escalatory control; accidental war; attribution; crisis instability; and defensive asymmetry. Let's look at each in turn.

1. DETERRENCE

Obviously, in this case deterrence failed. In our hypothetical scenario, the Cyber Command team was not deterred by considerations of what China might do to the U.S. In the real world, the U.S. probably should be deterred from initiating large-scale cyber warfare for fear of the asymmetrical effects that retaliation could have on American networks. Yet, deterrence is an undeveloped theoretical space in cyber war today. Deterrence theory was the underpinning of U.S., Soviet, and NATO nuclear strategy during the Cold War. The horror that could be caused by nuclear weapons (and the fear that *any* use would lead to *extensive* use) deterred nuclear-weapons nations from using their ultimate weapons against each other. It also deterred nations, both nuclear-armed and not, from doing anything that might provoke a nuclear response. Strategists developed complex theories about nuclear deterrence. Herman Kahn developed a typology with three distinct classes of nuclear deterrence in his works in the 1960s. His theories and analyses were widely studied by civilian and military leaders in both the United States and the Soviet Union. His clear, matter-of-fact writing about the likely scope of destruction in books like *On Thermonuclear War* (1960) and *Thinking About the Unthinkable* (1962) undoubtedly helped to deter nuclear war.

Of all the nuclear-strategy concepts, however, deterrence theory is probably the least transferable to cyber war. Indeed, deterrence in cyberspace is likely to have a very different meaning than it did in the works of Kahn and the 1960s strategists. Nuclear deterrence was based on the credible effects created by nuclear weapons. The world had seen what two nuclear weapons had done to Hiroshima and Nagasaki in 1945. Much larger nuclear weapons had then been detonated aboveground by the United States and the Soviet Union in the 1940s and 1950s, followed by the United Kingdom in 1952,

France in 1960, and China in 1968. All told, the initial five nuclear-weapons states detonated over 2,300 weapons above and below the surface.

No one knew exactly what would happen if either the United States or the Soviet Union tried to launch several hundred nuclear-armed ballistic missiles more or less simultaneously, but internally the American military thought that over 90 percent of its missiles would launch, make it to their targets, and detonate their weapons. They had similarly high expectations that they knew what the effects of their weapons would be on the targets. To insure a major attack would work, if attempted, the U.S. military planned on hitting important targets with nuclear warheads from three different delivery mechanisms (bombs from aircraft, warheads from ground-based missiles, and warheads from submarine-launched missiles). Both superpowers deployed their forces in such a way that they would have many surviving nuclear weapons even after suffering a large, surprise attack. Retaliation was assured. Thus, there was near certainty that by one side's using nuclear weapons, it was inviting some degree of its own nuclear destruction. What would happen after a massive exchange of nuclear weapons was subject to debate, but few doubted that the two nuclear combatants would have inflicted on each other a level of damage unparalleled in human history. Many believed a large-scale exchange would trigger a "nuclear winter" that could cause the end of all human life. Almost all experts believed that a large-scale exchange by the two superpowers would cause what were termed "prompt deaths" in the scores of millions. (Kahn dryly noted, "No one wants to be the first to kill a hundred million people.") Any use of nuclear weapons, it was feared, could escalate unpredictably into large-scale use. That fear has deterred the United States and the Soviet Union from using their nuclear weapons for over six decades to date.

The nuclear tests had created what was called a "demonstration effect." Some theorists also suggested that in a major crisis, such as a conventional war in Europe, the United States might detonate a nuclear weapon at sea as another demonstration effect, thus signaling that unless the fighting stopped, the NATO Alliance was prepared to escalate to nuclear-weapons use. NATO planned that during a conventional war it could "signal NATO's intent" by such a warning shot. Despite the instances of cyber war to date, the demonstration effect has not been compelling. As discussed earlier, most of the cyber incidents thus far have been either unsophisticated attacks such as a DDOS or covert penetrations of networks to steal information or implant trapdoors and logic bombs. The limited effects of the DDOS attacks were not widely noticed by those outside of the countries victimized. And in the case of most of the covert attacks, even the victims may not have noticed.

So what confidence do cyber warriors have that their weapons will work, and what expectation do they have about the effects that would be caused by the weapons? What they undoubtedly know is that they have already used many of their attack techniques to successfully penetrate other nations' networks. They have probably done everything short of a few keystrokes of what they would do in real cyber war. On simulations of enemy networks, they have probably engaged in destructive operations. The Aurora test on the generator in Idaho was one such test. It left the experimenters confident that they could have caused the physical destruction of a large electric generator with a cyber weapon.

What cyber warriors cannot know, however, is whether the nation they are targeting will surprise them with a significantly improved array of defenses in a crisis. What would be the effect if China disconnected its networks from international cyberspace? Would the U.S. plans for dealing with that contingency work? Assuming Russia has placed trapdoors and logic bombs in U.S. networks, how do they know whether the Americans have identified them and have planned their elimination in a period of heightened tension? When a cyber warrior goes to use the penetration technique he has planned on to get back into a target, that route of access may be blocked and an unexpected and effective intrusion-prevention system may suddenly have appeared. Unlike a national antimissile system, an intrusion-prevention for key networks could easily be kept secret until activated. If the cyber warrior's job is to shut down an enemy's air defense system slightly in advance of his nation's air force doing a bombing mission, the bombers may be in for a rude awakening. The radar installations and missiles that were supposed to have been shut down may suddenly come alive and destroy the attacking aircraft.

With a nuclear detonation, one could be fairly certain about what would happen to the target. If the target was a military base, it would become unusable for years, if not forever. On my first day of graduate school at MIT in the 1970s, I was given a circular slide rule, which was a nuclear-effect calculator. Spin one circle and you picked the nuclear yield, say 200 kilotons. Spin another circle and you could choose an airburst or a groundburst. Throw in how far away from the target you might be in a worst case and your handy little spinning device told you how many pounds of explosive pressure per square inch would be created and how many would be needed to collapse a hardened underground missile silo in on itself, before becoming little radioactive pieces of dust thrown way up in the atmosphere. A cyber warrior may possibly have similar certainty that were he to hit some system with a sophisticated cyber weapon, that system, say a modern freight railroad, would likely stop cold. What he may not know is whether the railroad has a reliable resiliency plan, a backup command-and-control network that he does not know about because the enemy is keeping it secret and not using it until it's needed. Just as a secret intrusion-prevention system might surprise us when it's suddenly turned on in a crisis, a secret continuity-of-operations system that could quickly get the target back up and running is also a form of defense against cyber attack.

The potential surprise capability of an opponent's defense makes deterrence in cyber war theory fundamentally different from deterrence theory in nuclear strategy. It was abundantly clear in nuclear strategy that there was an overwhelming case of what was called "offensive preference," that is to say, any defense deployed or even devised could easily be overwhelmed by a well-timed surprise attack. It costs far less to modify one's missile offense to deal with defensive measures than the huge costs necessary to achieve even minimally effective missile protection. Whatever the defense did, the offense won with little additional effort. In addition, no one thought for a moment that the Soviet Union or the United States could secretly develop and deploy an effective missile-defense system. Ronald Reagan hoped that by spending billions of dollars on research, the U.S. could change the equation and make strategic nuclear missile defense possible. Decades later it has not worked, and today the U.S. hopes, at best, to be able to stop a small missile attack launched by accident or a minor power's attack with primitive missiles. Even that remains doubtful.

In strategic nuclear war theory, the destructive power of the offense was well known, no defense could do much to stop it, the offense was feared, and nations were thereby deterred from using their own nuclear weapons or taking other provocative steps that might trigger a nuclear response. Deterrence derived from sufficient certainty. In the case of cyber war, the power of the offense is largely secret; defenses of some efficacy could possibly be created and might even appear suddenly in a crisis, but it is unlikely any nation is effectively deterred today from using its own cyber weapons in a crisis; and the potential of retaliation with cyber weapons probably does not yet deter any nation from pursuing whatever policy it has in mind.

Assume for the sake of discussion that the United States (or some other nation) had such powerful offensive cyber weapons that it could overcome any defense and inflict significant disruption and damage on some nation's military and economy. If the U.S. simply announced that it had that capability, but disclosed no details, many opponents would think that we were bluffing. Without details, without ever having seen U.S. cyber weapons in action, few would so fear what we could do as to be deterred from anything.

The U.S. could theoretically look for an opportunity to punish some bad actor nation with a cyber attack just to create a demonstration effect. (The U.S. used the F-117 Stealth fighter-bomber in the 1989 invasion of Panama not because it feared Panamanian air defenses, but because the Pentagon wanted to show off its new weapon to deter others. The invasion was code-named Operation Just Cause, and many in the Pentagon quipped that the F-117 was sent in "*just cause* we could.") The problem with the idea of using cyber weapons in the next crisis that comes up is that many sophisticated cyber attack techniques may be similar to the cryptologist's "onetime pad" in that they are designed for use only once. When the cyber attack weapons are used, potential opponents are likely to detect them and apply all of their research capability in coming up with a defense.

If the U.S. cannot deter others with its secret cyber weapons, is it possible that the U.S. itself may be deterred by the threat from other nations' cyber warriors? In other words, are we today self-deterred from conventional military operations because of our cyber war vulnerabilities? If a crisis developed in the South China Sea, as in the exercise described above, I doubt that today anyone around the table in the Situation Room would say to the President, "You better not send those aircraft carriers to get China to back down in that oil dispute. If you do that, Mr. President, Beijing could launch a cyber attack to crash our stock market, ground our airlines, halt our trains, and plunge our cities into a sustained blackout. There is nothing we have today that could stop them, sir."

Somebody *should* say that, because, of course, it's true. But would they? Very unlikely. The most senior American military officer just learned less than two years ago that his operational network could probably be taken down by a cyber attack. The Obama White House did not get around for a year to appointing a "cyber czar." America's warriors think of technology as the ace up their sleeves, something that lets their aircraft and ships and tanks operate better than any in the world. It comes hard to most of the U.S. military to think of technology as something that another nation could use effectively against us, especially when that technology is some geek's computer code and not a stealthy fighter-bomber.

So, we cannot deter other nations with our cyber weapons. In fact, other nations are so undeterred that they are regularly hacking into our networks. Nor are we likely to be deterred from doing things that might provoke others into making a major cyber attack. Deterrence is only a potential, something that we might create in the mind of possible cyber attackers if (and it is a huge if) we got serious about deploying effective defenses for some key networks. Since we have not even started to do that, deterrence theory, the sine qua non of strategic nuclear war prevention, plays no significant role in stopping cyber war today.

2. NO FIRST USE?

One of the first things you should have noticed about the scenario in our hypothetical exercise was the idea of *going first*. In the absence of any strategy to the contrary, the U.S. side in the hypothetical exercise took the first move in cyberspace by sending out an insulting e-mail on what China thought was its internal military e-mail system and then by initiating what the U.S. team hoped would be a limited power blackout. The strategic goal was to signal both the seriousness with which the U.S. viewed the crisis and the fact that the U.S. had some potent capabilities. Cyber Command's immediate tactical objective was to slow down the loading of the Chinese amphibious assault force, to buy time for U.S. diplomats to talk China out of its planned operation.

In nuclear war strategy, the Soviet Union proposed that we and they agree that neither side would be the first to use nuclear weapons in a conflict. The U.S. government never agreed to the No First Use Declaration, preserving for itself the option to use nuclear weapons to offset the superior conventional forces of the Soviet Union. (My onetime State Department colleague Jerry Kahan once asked a Soviet counterpart why they kept suggesting we ban orange juice. When the Russian denied making such a proposal, Jerry retorted, "But you're always running around saying 'no first juice.'") Should we incorporate a No First Use approach in our cyber war strategy?

There is no conventional military force in the world superior to that of the U.S., assuming that the U.S. military is not blinded or disconnected by a cyber attack. Therefore, we do not need to hold open the prospect of going first in cyberspace to compensate for some other deficiency, as we did in nuclear strategy. Going first in cyber war also makes it more politically acceptable in the eyes of the world for the victim of the cyber attack to retaliate in kind, and then some. Given our greater vulnerability to cyber attack, the U.S. may not want to provoke a cyber phase to a war.

However, forswearing the use of cyber weapons until they have been used on us could mean that if a conventional war broke out, we would not defend our forces by such things as cyber attacks on our opponent's antiaircraft missile systems. The initial use of cyber war in the South China Sea scenario was a psychological operation on China's internal military network, sending a harassing e-mail with a picture of a sinking Chinese ship. Should that be considered a first use of cyber war?

Moreover, the scenario presented a problem that if you do not go first in cyberspace, your ability to conduct cyber attack may be reduced by the other side stepping up both its defensive measures (for example, China cutting off its cyberspace from the rest of the world) and its offensive measures (including attacks that disrupted U.S. networks that may be necessary for some of the U.S. attacks to be launched). Whether we say it publicly or maintain it as an internal component of our strategy, if we were to accept the concept of No First Use in cyber war we would require a clear understanding of what constitutes "use." Is penetration of a network a cyber war act? When the network penetration goes beyond just collecting information, does the act then move from intelligence operations to cyber war? Any ban on "first use" would probably only apply prior to kinetic shooting. Once a war goes kinetic, most bets are off.

3. PREPARATION OF THE BATTLEFIELD

Another thing that you should have caught is that it appears that both sides had hacked into each other's systems well before the exercise began. In the real world, they probably have actually done just that. How much of this is done and who approves it is an issue to be reviewed when creating a strategy.

If CIA sends agents into a country to conduct a survey for possible future sabotage and they leave behind a cache of weapons and explosives, under U.S. law such activity is considered covert action and requires a Presidential finding and a formal notification of the two congressional intelligence committees. In recent years, the Pentagon has taken the view that if it conducts some kind of covert action, well, that's just preparation of the battlefield and no one needs to know. The phrase "preparation of the battlefield" has become somewhat elastic. The battle does not need to be imminent, and almost anyplace can be a battlefield someday.

This elasticity has also been applied to cyber war capability, and apparently not just by the United States. In the hypothetical exercise, both the U.S. and China opened previously installed trapdoors in the other country's networks and then set off logic bombs that had been implanted earlier in, among other places, the electric power grids. Beyond the exercise, there is good reason to believe that someone actually has already implanted logic bombs in the U.S. power grid control networks. Several people who should know implied or confirmed that the U.S. has also already engaged in the same kind of preparation of the battlefield.

Imagine if the FBI announced that it had arrested dozens of Chinese government agents running around the country strapping C4 explosive charges to those big, ugly high-tension transmission line towers and to some of those unmanned step-down electric substation transformers that dot the landscape. The nation would be in an outrage. Certain Congressmen would demand that we declare war, or at least slap punitive tariffs on Chinese imports. Somebody would insist that we start calling Chinese food "liberty snacks." Yet when the *Wall Street Journal* announced in a headline in April 2009 that China had planted logic bombs in the U.S. grid, there was little reaction. The difference in response is indicative mainly of the Congress, the media, and the public's inexperience with cyber war. It is not reflective of any real distinction between the effects those logic bombs could have on the power grid, compared to what little parcels of C4 explosives might do.

The implanting of logic bombs on networks such as the U.S. power grid cannot be justified as an intelligence-collection operation. A nation might collect intelligence on our weapon systems by hacking into Raytheon's or Boeing's network, but there is no informational value in being inside Florida Power and Light's control system. Even if there were valuable data on that network, logic bombs do not collect information, they destroy it. The only reason to hack into a power grid's controls, install a trapdoor so you can get back in quickly later on, and leave behind computer code that would, when activated, cause damage to the software (and even the hardware) of the network, is if you are planning a cyber war. It does not mean that you have already decided to conduct that war, but it certainly means that you want to be ready to do so.

Throughout much of the Cold War and even afterward there were urban legends about Soviet agents sneaking into the U.S. with small nuclear weapons, so-called suitcase bombs, that could wipe out U.S. cities even if Russian bombers and missiles were destroyed in some U.S. surprise attack. While both the Soviets and the U.S. did have small weapons (we actually had a few hundred called the Medium Atomic Demolition Munitions, or MADM, and another bunch called the Small Atomic Demolition Munitions, or SADM, which were designed to be carried in a backpack), there is no evidence that either side actually deployed them behind the other's lines. Even at the height of the Cold War, decision makers thought that actually sending the MADMs out onto the streets would be too destabilizing.

How is it, then, that Chinese, and presumably U.S., decision makers have authorized placing logic bombs on the territory of the other? It is at least possible that high-level officials in one or both countries never approved the deployments and do not know about them. The cyber weapons might have been implanted on the authority of military commanders acting under their authority to engage in preparation of the battlefield. There is a risk that senior policy makers will be told in a crisis that the other side has planted logic bombs in preparation for war and will view that as a new and threatening development, causing the senior policy makers to ratchet up their response in the crisis. Leaders may be told that since it is obvious the other side intends to crash our power grid, we should go first while we still can. Another risk is that the weapon may actually be used without senior-level approval, either by a rogue commander or by some hacker or disgruntled employee who discovers the weapon.

Cyber warriors justify the steps they have taken in preparation of the battlefield as necessary measures to provide national decision makers with options in a future crisis. "Would you want the President to have fewer courses of action to choose from in some crisis?" they would say. "If you want him to have the choice of a nonkinetic response in the future, you have to let us get into their networks now. Just because a network is vulnerable to unauthorized penetration now does not mean it will be so years from now when we may want to get in."

Networks are constantly being modified. An electric power transmission company might one day buy an effective intrusion-prevention system (IPS) that would detect and block the techniques we use to penetrate into the network. But if we can get into the network now, we can leave behind a trapdoor that would appear to any future security system as an authorized entry. Getting onto the network in the future is not enough, however; we want to be able to run code that makes the system do what we want, to malfunction. That future IPS might block the downloading of executable code, even by an authorized user, without some higher level of approval. Thus, if we can get into the system now, we should leave behind the instruction code to override surge protection or cause the generators to spin out of synchronization, or whatever method we have to disrupt or destroy the network or the hardware it runs.

That sounds persuasive at one level, but are there places where we do not want our cyber warriors preparing the battlefield?

4. GLOBAL WAR

In our hypothetical exercise, the Chinese response aimed at four U.S. navy facilities but spilled over into several major cities in four countries. (The North American Interconnects link electric power systems in the U.S., Canada, and parts of Mexico.)

To hide its tracks, the U.S., in this scenario, attacked the Chinese power grid from a computer in Estonia. To get to China from Estonia, the U.S. attack packets would have had to traverse several countries, including Russia. To discover the source of the attacks on them, the Chinese would probably have hacked into the Russian routers from which the last packets came. In response, China hit back at Estonia to make the point that nations that allow cyber attacks to originate from their networks may end up getting punished even though they had not intentionally originated the attack.

Even in an age of intercontinental missiles and aircraft, cyber war moves faster and crosses borders more easily than any form of hostilities in history. Once a nation-state has initiated cyber war, there is a high potential that other nations will be drawn in, as the attackers try to hide both their identities and the routes taken by their attacks. Launching an attack from Estonian sites would be like the U.S. landing attack aircraft in Mongolia without asking for permission, and then, having refueled, taking off and bombing China. Because some attack tools, such as worms, once launched into cyberspace can spread globally in minutes, there is the possibility of collateral damage as these malicious programs jump international boundaries and affect unintended targets. But what about collateral damage in the country that is being targeted?

5. COLLATERAL DAMAGE AND THE WITHHOLD
DOCTRINE

Trying to strike at navy bases, the two cyber combatants hit the power plants providing the bases electricity. In so doing, they left large regions and scores of millions of people in the dark because electric power grids are extremely vulnerable to cascading failures that move in seconds. In such a scenario there would probably be dozens of hospitals whose backup generators failed to start. The international laws of war prohibit targeting hospitals and civilian targets in general, but it is impossible to target a power grid without hitting civilian facilities. In the last U.S.-Iraq War, the U.S. campaign of "Shock and Awe" employed precision-guided munitions that wiped out targeted buildings and left structures across the street still standing. While being careful with bombs, the U.S. and other nations have developed cyber war weapons that have the potential to be indiscriminate in their attacks.

In the cyber war game scenario, U.S. Cyber Command was denied permission to attack the banking sector. In the real world, my own attempts to have NSA hack into banks to find and steal al Qaeda's funds were repeatedly blocked by the leadership of the U.S. Treasury Department in the Clinton Administration. Even in the Bush Administration, Treasury was able to block a proposed hacking attack on Saddam Hussein's banks at the very time that the administration was preparing an invasion and occupation in which over 100,000 Iraqis were killed. Bankers have successfully argued that their international finance and trading system depends upon a certain level of trust.

The U.S. decision to withhold attacks narrowly targeted on the financial sector also reflects an understanding that the United States might be the biggest loser in a cyber war aimed at banks. Even though the financial services sector is probably the most secure of all of the major industry verticals in the U.S., it is still vulnerable. "We've tested the security at more than a dozen top U.S. financial institutions, as hired consultants, and we've been able to hack in every time," one private-sector security consultant told me. "And every time, we could have changed numbers around and moved money, but we didn't."

The existing U.S. policy does not prohibit hacking into foreign banks to collect intelligence, but it does create a very high hurdle for altering data. Both the Secretary of the Treasury and the Secretary of State have to personally authorize such an action. As far as I was able to determine from my sources, that approval has never been granted. We have, in effect, what in nuclear war strategy we called a "withhold target set," things that we have targeted but do not intend to hit. That policy assumes, or hopes, that opponents will also play by those unarticulated rules. In Exercise South China Sea, the PLA team did not. In its last move it hit the databases of the stock market and the major bank clearing house. That was a dramatic and, we hope, unrealistic escalation. Today China's economy is so tightly connected to America's that they, too, might have a withhold doctrine affecting the financial sector. Under foreseeable circumstances, it is maybe an acceptable risk to assume that nations will all withhold data-altering attacks on the financial sector, though some U.S. analysts would dispute that about China.

Because a sophisticated nonstate actor might not be so polite, it would be important for the U.S. financial sector to have an advance understanding with the federal regulators about what they would do if there were a major hack that altered data. Certain European and Japanese institutions should probably also be discreetly consulted about the policies they would use to reconstruct who owns what after a major data-altering breach. The Federal Reserve Bank and the Securities Industry Automation Corporation, among other financial database operators, have extensive off-site backup systems. Key to their being prepared to fix a data-altering breach is the idea that there is data with a recent picture of "who owns what" that is unlikely to be altered by a cyber attack. With the agreement of the federal regulators, banks and stock markets could revert to a prior date to recover from a data-altering breach. Some people would be hurt and others enriched by such a decision and it would be the subject of litigation forever, but at least the financial system could continue to operate.

China's air traffic control (ATC) system was also placed on a withhold list in the exercise. As the U.S. modernizes its ATC, making it more network dependent, the system is likely to become only more vulnerable to cyber attack. Already with the older system, the U.S. has experienced instances where individual airport towers and even specific regional centers have been blacked out for hours because of computer or communications connectivity failures. As far as we know, none of these major outages was caused by hacking. (There is one case of an arrest for hacking into the FAA system, but the effects of the attack were minor.)

Nonetheless, the potential for someone altering data and causing aircraft to collide in midair has to be considered. The U.S. is a party to the Montreal Convention, which makes an intentional attack on a civilian airliner a violation of international law. Of course, almost all hacking is a violation of some national and/or international law, but the Montreal Convention is an articulation of the general global sentiment that certain kinds of actions are beyond the pale of acceptable conduct.

Hacking into the flight controls of an aircraft in flight is probably also becoming more feasible. The Federal Aviation Agency raised concerns with Boeing that plans for the new 787 Dreamliner called for the flight control system and the elaborate interactive passenger-entertainment system to use the same computer network. The FAA was concerned that a passenger could hack into the flight control system from his seat, or that live Internet connectivity for passengers could mean that someone on the ground could hack into the system. The airlines' own systems already create a data connection from the ground to some aircraft's computer networks. The computer networks on a large passenger aircraft are extensive and play a significant role in keeping the aircraft in the air.

In modern "fly-by-wire" aircraft, it is the flight control system that sends a computer signal to a flap, aileron, or rudder. The Air France crash over the South Atlantic in 2009, mentioned earlier, revealed to a wider audience what pilots have known for years: in modern fly-by-wire aircraft, onboard computers decide what signals to send to the control surfaces. Under certain circumstances, the software can even override the decision of a pilot to prevent the manual controls from making the aircraft do something that would cause it to stall or go out of control. As that recent Air France crash also demonstrated, the aircraft's computers were firing off messages back to the Air France headquarters' computers without the pilot being involved. As with the ATC system, the computer networks of commercial passenger aircraft should probably also be off limits. Military aircraft are, however, likely to be considered fair game.

Had the Cyber Command team asked the Controllers for permission to attack the reservations and operations network of Chinese airlines, they may have gotten a different answer. In the real world, computer crashes at U.S. and Canadian airlines have kept hundreds of aircraft grounded for hours at a time. The aircraft worked and there were crews available, but without the reservations database and the operational network up and running, the airlines did not know what crew, passengers, cargo, or fuel load should go on what planes. The airlines, like so many other huge business systems, no longer have manual backup systems that are sufficient to create even minimal operations.

There may be other withholds, in addition to banking and commercial aircraft. In the exercise, two of the networks that Cyber Command was told not to strike were China's military command and control network and their air defense system. Since those are purely military targets, why were they spared?

6. ESCALATORY CONTROL

During the Cold War, I often participated in exercises in which teams of national security officials were secretly hustled out of Washington on short notice to obscure, covert locations. Once at our destinations, the teams did exactly what the *War Games* movie computer suggested. We played thermonuclear war. These were massively depressing experiences, since the "game reality" we had to accept was that millions of people had already died in a nuclear exchange. Our job was almost always to finish the war and begin the recovery.

The most difficult part of finishing the war usually turned out to be finding who was still alive and in control of the military on the other side. What survivor was in command of Soviet forces, and how do we talk to him without either of us revealing our hidden locations? Part of the problem that the game controllers deviously planned for us sometimes was that the guy with whom we were negotiating war termination did not actually have control over some element of the surviving Soviet force, for instance, their nuclear missile submarines. What we learned from these unpleasant experiences was that if we eliminate the opponent's command and control system, then he has no way to tell his forces to stop fighting. Isolated local commanders, cut off from communications with higher echelons, or not recognizing the authority of the surviving successor, made their own decisions, and often it was to keep fighting. It was the nuclear war equivalent of those lone Japanese fighters who kept turning up on remote Pacific isles in the 1950s, unaware that the Emperor had years before ordered them to surrender.

There may be a parallel in cyber war. If a cyber attack eliminates a military command and control system, it could be difficult to prevent or terminate a kinetic war. In most militaries authority devolves to the local commander if he cannot get in touch with his superiors. Even if the command system is still operating, if the local commander believes that the system has been taken over by an opponent who is now issuing false instructions, command probably devolves to the local general until he can ascertain that he is in reliable communications with a valid superior. This is the situation so vividly portrayed in the movie *Crimson Tide*, where the U.S. nuclear submarine commander received and authenticated an order to launch nuclear missiles and then received an order to stop. Unable to authenticate the order to stop the attack, and fearing that it was a bogus order somehow sent by the Russians, the captain believes that procedures require him to launch.

The conclusion that we came to repeatedly in the nuclear war games was that it was probably an error to engage in a "decapitating strike," one that made it impossible for the leadership to communicate with us or with their own forces. In cyber war, it may be desirable to cut off certain units from higher command, or to deny an opposing force access to intelligence about what is going on. But in choosing what units to cut off, one needs to keep in mind that severing the command link to a unit runs the risk that it will launch an attack on its own. Thus, cyber attacks should probably be carefully constructed so that there is still a surviving communications channel for negotiations and a way in which the leadership can authoritatively order its forces to stop fighting.

The exercise's Control Team also denied Cyber Command the authority to strike at air defense networks. The rationale for that kind of withhold at that point in hostilities is "escalatory control."

In his 1965 masterpiece of military strategy, *On Escalation*, Kahn argued that if your goal is war termination short of the total destruction or forced surrender of the opponent, you can signal that by what you strike and what you withhold. You may want to signal that you have limited intentions so that the other side does not assume otherwise and proceed as if it has nothing to lose.

There are cyber war corollaries to escalation control. A cyber attack on a nation's air defense system would lead that country's leadership to the logical conclusion that air attacks were about to happen. In Exercise South China Sea, there were U.S. aircraft carriers nearby. If the Chinese military thought that those carriers were getting ready for air strikes on China, they would be right to take preemptive steps to sink the carriers. So, a cyber attack on the air defense network could have caused the beginning of a kinetic war that we were seeking to avoid. Even an attempted penetration of that network to lay in trapdoors and logic bombs might have been detected and interpreted as a prelude to imminent bombing. So just getting into position to launch a cyber attack would have sent the wrong message in a crisis, unless those steps had been taken well in advance.

Herman Kahn, Thomas Schelling, William Kaufmann, and the other "Wizards of Armageddon" spent a lot of time thinking about how to control nuclear escalation, from the tensions leading up to a crisis, to signaling, to initial use, to war termination. Initially the nuclear strategists saw war moving slowly up the escalatory ladder, with diplomatic attempts being made at every rung to stop the conflict right there. They also discussed what I referred to earlier, "escalation dominance." In that strategy, one side says, basically, "We don't want to play around with low-grade fighting that will gradually get bigger. If you want to fight me, it's going to be a big, damaging fight." It's the warfare equivalent of going all-in on a hand in poker and hoping your opponent will give up rather than risk all of his chips. Except that there is one big difference:

in escalation dominance you are actually jumping several rungs up the ladder and inflicting serious damage on the other side. You accompany that move with the threat that you can and will do more significant damage unless it all stops right here, right now.

The fact that you have done that damage to them may cause the opponent to feel compelled to respond in kind. Or, if you have a highly rational actor on the other side, they'll understand that the stakes are getting too high and they stand to suffer even more serious losses if things continue. In Exercise South China Sea, the PLA decided to engage in escalation dominance. In response to a cyber attack on the power grid in southeastern China, they not only hit the West Coast power grid, they disrupted the global Defense Department intranet, damaged the databases of U.S. financial clearinghouses, and sent additional kinetic warfare units into the crisis zone in the South China Sea.

As the game continued, the U.S. leadership had to decide quickly whether it stood to lose more than China in the next round of cyber war escalation. America would have been at a disadvantage, because it stood to lose more in an ongoing, escalating cyber war. It therefore sought a quick diplomatic settlement. Escalation dominance was the right move for China in this game because that escalation showed that the U.S. was more susceptible to cyber attacks and that further escalation would only make matters worse for the U.S. team. The U.S. could have tried to block cyber traffic coming from China. But because the Chinese attacks were originating inside the U.S., and there was not yet a deep-packet inspection system on the Internet backbone, the next, larger, Chinese cyber attack would have been very difficult to stop.

Put more simply, if you are going to throw cyber rocks, you had better be sure that the house you live in has less glass than the other guy's, or that yours has bulletproof windows.

7. POSITIVE CONTROL AND ACCIDENTAL WAR

The issue we discussed above, of maintaining some means for the opponent to exercise command and control, raises a similar issue, namely: Who has the authority to penetrate networks and to use cyber weapons? Earlier in this chapter I suggested that it may require the approval of multiple Cabinet members to alter banking data, and yet we are not sure that the President knows that the U.S. may have placed logic bombs in various nations' power grids. Those two facts suggest that there is too much ambiguity regarding who has what authority when it comes to cyber war, including preparation of the battlefield.

In nuclear war strategy there were two central issues regarding who could do what, and they came under the general heading of "positive control." The first was, simply: Could some U.S. military officer who had a nuclear weapon use that weapon even if he was not authorized to do so? To prevent that from happening, as well as to prevent someone from stealing and then setting off a bomb, elaborate electronics were embedded in the bomb's design. The electronics physically blocked the bomb from detonating unless the lock had received an alphanumeric unlocking code. On many weapons, two officers had to each confirm the code and simultaneously turn physical keys to accomplish their part of the unlocking sequence. This was called the "two key" control. Part of that code was kept away from the weapon and would be sent down by higher authority to those who would unlock it. These "permissive action links," or PALs, grew more sophisticated over the years. The U.S. shared parts of its PAL technology with some other nuclear-weapon states.

The second issue regarding positive control was: Who should be the higher authority capable of sending down the unlocking codes for nuclear weapons? The theory was that under normal circumstances that authority would rest with the President. A military officer attached to the U.S. President carries at all times a locked case in which reside the "go codes" for various nuclear attack options. I learned during the attempted military coup in Moscow in 1990 that the Soviets had a similar system. President Gorbachev, who was taken hostage at one point in the crisis, had the nuclear "go codes" with him at his vacation villa. The Gorbachev incident highlights the need for having the decision-making authority devolve if the President is unable to act. The U.S. government refuses to acknowledge who below the President has the authority to unlock and use nuclear weapons and under what circumstances that power devolves. All personnel who have physical access to nuclear weapons must undergo special security review and testing as part of a personnel reliability system designed to weed out persons with psychological or emotional issues.

Cyber weapons would have a far lesser impact than nuclear weapons, but their employment under certain circumstances could be highly damaging and could also trigger broader war. So, who gets to decide to use them, and how do we make sure they are not used without authorization? Who should decide what networks we should be penetrating as part of the preparation of the battlefield?

Until we gain more experience with cyber weapons, I would argue that the President should at least annually approve broad guidelines about what kinds of networks in what countries we should be penetrating for both intelligence collection and for the embedding of logic bombs. Some will criticize that as overly restrictive, noting that we have been penetrating networks for intelligence collection for years without presidential review. That may be true, but in many cases there are only a few keystrokes' difference between penetrating a network to collect intelligence and hacking your way in to cause destruction and disruption. Because there is the risk, however low, that logic bombs and other penetrations may be discovered and misunderstood as hostile intentions, the President should decide on how much risk he wants to take, and with whom.

The decision to use a cyber weapon for disruptive or destructive purposes should also rest with the President, or, in rare cases where quick action is necessary, with the Secretary of Defense. There may be circumstances in which regional commanders should have some predelegated authority to respond defensively to an ongoing or imminent attack. However, Cyber Command and its subordinate units should employ some form of software control analogous to the two-key control on nuclear weapons to ensure that an overzealous or massively bored young lieutenant cannot initiate an attack.

Even with proper command controls in effect, there is the potential for accidental war. In the Cold War, early radar systems could sometimes not distinguish between huge flocks of Canada geese and formations of Russian bombers. Thus, there were times when the U.S. launched the portion of its bombers that were kept on strip alert and sent them heading toward their destinations until air defense authorities could clarify the situation and determine for sure if we were under attack.

In cyber war, it is possible to imagine accidental attacks developing if somehow the wrong application were used and instead of inserting code that copied data, we mistakenly used code that deleted data. Alternatively, you could imagine the possibility that a logic bomb might be accidentally triggered by the network operator or by some other hacker who found it. The chances of that happening are very low, but Cyber Command and others engaged in hacking into other nations' networks must have strict procedures to ensure that no such mistake occurs. The greatest potential for accidental cyber war is likely to come in the form of retaliating against the wrong nation because we were misled as to who attacked us.

8. ATTRIBUTION

In Exercise South China Sea, neither side doubted the identity of who was attacking them. There was a political context, rising tensions over the offshore oil fields. But what if, instead of China having done the attack, it was Vietnam? In the exercise scenario, Vietnam and the U.S. are allied against China. So why would our ally attack us? Perhaps Vietnam wants to drag the U.S. deeper into the conflict, to get Washington to stand up against China. What better way than letting Washington think that China was engaged in cyber war against us? And when China denied that it was them, we would probably just write that off as Beijing engaging in plausible deniability. (If you want to contemplate a similar scenario, and if you will forgive a bit of shameless self-promotion, read my novel *Breakpoint*, which deals with cyber war attribution, among other things.)

The cyber experts at Black Hat were asked at the 2009 meeting whether they thought the problem of attribution was as important as some suggest, that is, is it really that hard to figure out who is attacking you, and does knowing who attacked you really matter? To a person, they answered that attribution was not a major issue to them. It was not that they thought it was easy to identify the attacker; rather, they just did not care who it was. These were mainly corporate people whose networks had been attacked and when it had happened, their chief concern was getting the system back to normal and preventing that kind of attack from happening again. Their experiences dealing with the FBI had convinced most of them that it was hardly worth it even to report to law enforcement when they had been attacked.

For national security officials, however, knowing who attacked you is much more important. The President may ask. You may want to send the attacker a diplomatic note of protest, a demarche (what we called in the State Department a "démarche-mallow."), as Secretary Clinton did after news of the attempted hacking on Google from Mainland China went public. You might even want to retaliate to get them to stop doing it. One way to find out who the attacker was is to use trace-back software, but eventually you will probably get to a server that does not cooperate. You could, at that point, file a diplomatic note requesting that the law enforcement authorities in the country get a warrant, go around to the server, and pull its records as part of international cooperation in investigating a crime. That could take days, and the records might be destroyed by then. Or the country in question may not want to help you. When trace-back stops working, you do have the option of "hack back," breaking into the server and checking its records. Of course, that is illegal for U.S. citizens to do, unless they are U.S. intelligence officers.

Hacking into a server to trace the origin of an attack may not work, either, if the attacker worked hard at covering up his origins. You may have to be online, watching live when the attack packets actually move through the servers. It is unlikely that you will find that, say, even after bouncing through a dozen servers in as many countries to cover their tracks, the attacking packets had originated in some place called the "Russian Offensive Cyber War Agency." Just to be safe, if it were the Russian government, they probably would have directed the attack from a server in another country and, if it were an intelligence-collection operation, the data they copied would probably have been sent to a data-storage unit in a third country.

So when it comes to figuring out who attacked you, unless you are sitting on the network the attacker uses and you see it coming (and sometimes not even then), you may not know right away. Computer forensics may be able to say that the original keyboard used in developing the attack code was designed for Arabic, or Cyrillic, or Korean, but that is hardly dispositive as to the identity of the hacker. And if you do find that the attack came from Russia, based on what happened to Estonia and Georgia, the authorities there will likely blame citizen hacktivists and do nothing to them.

This attribution difficulty could mean that nations trying to identify their attackers may need to rely upon more traditional intelligence techniques, such as spies penetrating the other side's organization, or police methods. Human intelligence, unlike cyber, does not move at velocities approaching the speed of light. Quick responses may not be available. In nuclear war strategy, attribution was not generally thought to be a major problem because we could tell where a missile or bomber had been launched. Cyber attack may be similar to a suitcase bomb going off in an American city. If we see the attack being launched because we are watching the cyber equivalent of their missile silos and bomber bases, we might be able to assign attack attribution with a high degree of certainty. But if the attack starts on servers in the U.S., it may take a while to tell the President that we really know who attacked us. How sure do you need to be before you respond? The answer will likely depend upon the real-world circumstances at the time.

9. CRISIS INSTABILITY

The late Bill Kaufmann once asked me to write a paper on something called "launch on warning." The Strategic Air Command had the idea that as soon as we saw a Soviet nuclear attack coming we should launch as many bombers as we could and fire our land-based missiles. As the Soviets had improved the accuracy of their missiles, it had become possible for them to destroy our missiles even though we kept them in hardened, underground silos. As with everything in strategic nuclear doctrine, even this idea of "fire when you see them coming" got complicated. What if you were wrong, if your sensors made a mistake? Perhaps they were attacking, but with a small force aimed at only a few things, should you still throw the kitchen sink at them? Therefore the Air Force had evolved a strategy called "launch under attack," which essentially meant that you waited until you had a better picture, until some of their missiles' warheads were already going off in your countryside.

The launch on warning strategy was generally thought to be risky because it added to crisis instability, the hair-trigger phenomenon in a period of rising tensions. If you don't make the right decision quickly, you lose, but if you have to make the decision quickly, you may make a losing decision. What I was able to conclude for Kaufmann was that we had enough missiles at sea, and those missiles had grown sufficiently accurate, that we could ride out an attack and then make a rational decision about what had just happened before we sized our response.

There is a similar issue with cyber war. The U.S. expects to see an attack coming and move quickly to blunt the cyber assault and destroy the attacker's ability to try it again. The assumption about being able to see an attack coming may be invalid. Nonetheless, we will assume that the U.S. strategy is to see the attack coming and act. To act, you have to go quickly and without a lot of assessment of who the enemy was or what they were going to strike. If you do not go quickly, however, you suffer two possible disadvantages:

- The attacking nation will probably pull up the drawbridge over the moat after its attackers charge out of the castle, by which we mean that as soon as they launch a big attack, a nation like China may disconnect from the rest of the Internet and "island" subnets;
- The attacking nation may be going after the Internet itself and the telephone infrastructure in the United States, which might make it harder for the U.S. to launch a cyber retaliation.

Thus, there could be a real case of *first mover advantage*, and that leads to crisis instability, a hair trigger, no time to think. Now, remember the earlier discussion about *ambiguity of intent*, what one side indicates by the types of targets it goes after in the preparation-of-the-battlefield period. If a nation believes that the other side has already laced its infrastructure (including cyber and electrical networks) with destructive software packages or logic bombs, that consideration, combined with the first mover advantage, could cause a decision maker in a time of rising tensions to have a very itchy keyboard finger.

10. DEFENSIVE ASYMMETRY

The team playing China won this exercise, forcing a withdrawal of U.S. forces and causing the United States to negotiate a face-saving way out. The chief reason they won was that they had been able to overcome U.S. defenses and to erect relatively effective defenses of their own. The U.S. was looking for an attack to originate overseas, and China used servers in the U.S., perhaps directed by Chinese "students" operating out of coffee shops. The U.S. was looking for the signatures of attacks that it already knew about and the Chinese used "zero day" exploits. Most important, the U.S. had no national defense mechanism for the civilian infrastructure, including the finance industry, the electric power grid, and rail systems.

China, on the other had, not only had a national command system that could dictate to its infrastructure, they had a defensive plan. When it was clear that cyber war was under way, China's electric and rail systems shifted to a non-networked control system. When the Chinese lost satellite communications, they had a backup radio network up in an hour. In short, China had not thrown out their old systems, and had a plan to use them.

. . .

The lessons learned in the "hot wash" of this exercise have helped to identify issues and choices, which will lead us toward a cyber war strategy. There is, however, one further missing ingredient. We have talked a little about the international laws of war and other conventions. What international laws cover cyber war, and what additional multilateral agreements would be in our interest, if any?

CHAPTER SEVEN

CYBER PEACE

The United States, almost single-handedly, is blocking arms control in cyberspace. Russia, somewhat ironically, is the leading advocate. Given the potential destabilizing nature and disadvantages of cyber war to the U.S., as discussed in the earlier chapters, one might think that by now the United States would have begun negotiating international arms control agreements that could limit the risks. In fact, since the Clinton Administration first rejected a Russian proposal, the U.S. has been a consistent opponent of cyber arms control.

Or, to be completely frank, perhaps I should admit that I rejected the Russian proposal. There were many who joined me; few U.S. government decisions are ever the responsibility of a single person. However, one of my jobs in the Clinton White House was to co-ordinate cyber security policy, including international agreements, across the government. Despite some interest in the State Department in pursuing cyber arms control, and although the U.S. had to stand almost alone in the U.N. in rejecting cyber talks, we said no. I viewed the Russian proposal as largely a propaganda tool, as so many of their multilateral arms control initiatives had been for decades. Verification of any cyber agreement seemed impossible. Moreover, the U.S. had not yet explored what it wanted to do in the area of cyber war. It was not obvious then whether or not cyber war added to or subtracted from U.S. national security. So we said no, and we have kept saying no for over a decade now.

Now that over twenty nations' militaries and intelligence services have created offensive cyber war units and we have gained a better understanding of what cyber war could look like, it may be time for the United States to review its position on cyber arms control and ask whether there is anything beneficial that could be achieved through an international agreement.

A SHORT CRITIQUE OF ARMS CONTROL

Whether or not you think reviewing our position on cyber arms limitations is a good policy may well depend upon what you think about arms control more broadly. So let's begin by recalling what arms control is (since it no longer dominates the news) and what it has done in other areas. Although there were international arms control agreements before the nuclear era, such as the Washington Treaty that limited the number of battleships navies could have before World War II, arms control as we now know it was shaped by the Cold War standoff between the U.S. and the U.S.S.R. Beginning in the early 1960s and continuing for almost thirty years, arms control became a major preoccupation of the two nuclear superpowers. What resulted were two classes of agreements: multilateral treaties, in which the two superpowers invited global participation, and bilateral agreements, in which they agreed to impose specific limitations on their own military capabilities.

I began working on arms control in Vienna in 1974 and, at the Pentagon and then the State Department, was involved for almost twenty years in agreements on strategic nuclear weapons, conventional forces in Europe, so-called theater nuclear weapons of shorter range, biological weapons, and chemical weapons. That experience shapes the way I think about cyber arms control. There are lessons the United States can learn from this history as we seek to limit warfare in cyberspace through a new round of treaties.

My colleague Charles Duelfer, who was one of the leaders of UN efforts to limit Iraqi weapons of mass destruction for over a decade, takes a cynical view of U.S.-Soviet arms control and of the phenomenon in general. "The U.S. and U.S.S.R. generally agreed to ban things they were not going to do anyway. On weapons they did want, they agreed to numeric ceilings that were so high that they got to do everything they wanted." Many analysts, like Duelfer, have a negative critique of arms control in general. They note that the fifteen-year-long talks on forces in Central Europe finally produced an accord with high limits on military personnel only months before the Soviet Union's military alliance crumbled anyway. The final treaty allowed the Soviet Union to keep hundreds of thousands of troops in Eastern Europe, but reality did not. What caused the thousands of Red Army tanks to clank back into Russia was not arms control.

The more well known series of negotiations of the SALT and START agreements on strategic nuclear forces lasted over twenty years and permitted both sides to maintain enormous numbers of nuclear weapons and to continue to replace them with more modern versions. As part of that process, in the ABM Treaty, the two nations banned antiballistic missile defenses, which at the time neither side thought would work anyway.

In the multilateral arena, the two superpowers agreed on a treaty to prohibit other nations from acquiring nuclear weapons in exchange for a vague promise that the nuclear powers would eventually eliminate their own. That treaty did not stop Israel, Pakistan, India, South Africa, or North Korea from developing nuclear weapons and is now doing little to stop Iran. The Soviet Union agreed to a multilateral ban on biological weapons, but then secretly went on to create a massive biological weapons arsenal that the United States did not detect for decades. The critics of arms control point to the Soviet violation of the Biological Weapons Treaty as an example of why arms control is often not in the U.S. interest. The U.S. is fairly scrupulous in its obedience of treaty limits to which it agrees. Many other nations are not. Verification measures may not detect violations, or permitted activities may allow nations to come right up to the point of a violation without being sanctioned (as Iran may be doing with its nuclear reprocessing program).

For all the problems with arms control, there is a compelling case that both the bilateral agreements between the U.S. and U.S.S.R. and the broader multilateral treaties made the world safer. Even putting aside the value of the numeric limits on weapons, the very existence of a forum where the American and Soviet diplomats and military leaders could talk to each other about nuclear war helped to create a consensus among the elites of both countries to take measures to prevent such a disaster. The introduction of communications channels and confidence-building measures, the increase in transparency of both sides' armed forces reduced the possibility of miscalculation or accidental war.

As Assistant Secretary of State, it was my duty to supervise one of those so-called confidence-building measures, the U.S. Nuclear Risk Reduction Center. My counterpart was a Russian General in the Ministry of Defense. Our two teams worked on measures to reduce the likelihood of tensions escalating into nuclear alerts. Each team had a center, mine in the State Department and the general's in the Ministry of Defense, just off Red Square in Moscow. Because the White House–Kremlin hotline was seldom employed by U.S. Presidents, we needed a way of communicating quickly at a lower level when there may have been a misunderstanding. So we connected the two centers by direct cable and satellite links, by Teletype for text, and by secure telephones. The secure telephone had to use an encryption code that we and the Soviets could share, which posed a problem for both countries. We both wanted to use encryption that would provide no clue about codes either side used elsewhere. Such was the fear of electronic espionage that some people thought that with such connectivity, I was just providing a way for the Soviets to listen in on U.S. communications. The entire U.S. Center, just off the State Department Operations area, had to be lined in copper and acoustic dampening materials.

The Nuclear Risk Reduction Centers were designed to prevent the kind of mistaken escalation that occurred in the early days of the Cold War. One day when a U.S. space launch from an aircraft platform aborted, we realized that on Russian radar the descending missile could look like a single depressed-trajectory surprise attack, possibly aimed at decapitating the leadership by hitting Moscow. I quickly called my counterpart in the Defense Ministry, on the secure line. Those lines were used repeatedly in instances like that, as well as to coordinate implementation of arms control agreements.

While it is true that SALT and START permitted large arsenals to continue for a long time, the treaties did ban destabilizing activities and programs that both sides might otherwise have felt the need to test or deploy. The numeric limits also provided a known quantity to the other side's force, preventing an even greater upward arms spiral based on false assumptions about what the other was intending. Eventually, thanks to the persistence of National Security Advisor Brent Scowcroft, the two sides banned the highly destabilizing multiple-warhead land-based missiles. Now, the U.S. and Russia are making meaningful reductions in their strategic forces.

The Intermediate Nuclear Forces (INF) treaty, on which I worked for several years in the early 1980s, caused the United States to destroy its Pershing II mobile ballistic system and its ground-launched cruise missiles, or GLCMs (they were originally called land-launched cruise missiles, or LLCMs, until the way that acronym was pronounced—lickems—occasioned so many off-color jokes that the Pentagon changed it), in exchange for the destruction of hundreds of Soviet SS-4, SS-5, and SS-20 mobile nuclear missiles. That entire class of weapon, which could be used to circumvent limits on longer-range systems, was permanently banned and several thousand nuclear warheads in Europe were taken out of service.

The limits on nuclear weapons testing did begin with the modest prohibition of detonating weapons in the atmosphere, but over time evolved into a limit on the size of all nuclear tests and eventually to a ban on nuclear testing altogether. (The complete ban on testing has not yet been ratified by the U.S. Senate.) The ban on chemical weapons, which I worked on in the early 1990s, is causing nations to destroy their chemical weapons, prohibits making new ones, and has a very intrusive inspection regime for verification. (While we did not agree to "anytime, anywhere" inspection, few areas are exempt.)

Beyond the limits and bans on nuclear, chemical, and biological weapons, arms control includes limits on the conduct of war itself. A series of agreements on armed conflict bans attacks on military hospitals, prohibits attacks on civilian population centers, establishes standards for treating prisoners of war, bans torture, outlaws land mines, limits the use of child soldiers, and makes genocide an international crime. The United States has not ratified all of these agreements (such as the ban on land mines) and has recently violated others (such as the Convention Against Torture). World War II saw broad violations of the laws of armed conflict, but even then some nations upheld the standards for treatment of prisoners of war.

When arms control works well, it reduces uncertainty, creating a more predictable security environment. By establishing some practices as illegal and some armament acquisition as a violation, arms control agreements can clarify what another nation's intentions might be. If a nation is willing to violate a clear agreement, there is less ambiguity about their policies. By prohibiting certain arms and practices, arms control can sometimes help nations to avoid expenditures that they might have been driven to only by fear that other nations were about to do the same. Agreed-upon international norms can be useful in gathering multilateral support against a nation that is an outlier.

When arms control is not valuable and can even be unhelpful is when it is largely hortatory, or when the negotiation is seen as an end in itself or a platform for propaganda, when its limitations are vague and also when violations are without cost to the violator. If a nation can quickly move from compliance to significant violation with little or no warning time, the attributes of stability and predictability are lost. Similarly, if nations can cheat on agreements with little or no risk of detection or fear of punishment when caught, the agreements tend to be one-sided and are discredited.

My overall view is that the arms control experience we had in the last thirty years of the Cold War was largely positive, but it was very far from a panacea and occasionally it was little more than a farce. A simple test of whether an area is ripe for arms control is to determine if all parties have a real interest in limiting their own investments in the area. If a party is proposing to stop something that they really want to keep around, then they are likely merely engaged in arms control for propaganda or as a deceptive means of constraining a potential opponent in an area where they think they may be outclassed.

LIMIT CYBER WAR?

All of which brings us back to cyber war. To determine our national policy toward concepts of arms control or limits on cyber war activities, we first need to ask whether this new form of combat gives the United States such an advantage over other nations that we would not wish to see international constraints. If we believe that we do enjoy such a unilateral advantage, and that it is likely to continue, then we should not ask the follow-on questions about what kinds of limits might be created, whether they could be verified, and so on.

I suggested earlier that at present the U.S. would be better off if cyber warfare never existed, given our asymmetrical vulnerabilities to such warfare. Before looking at cyber war control, let's first consider four ways in which we are more vulnerable than those nations that might use cyber weapons against us. First, at the moment, the United States has a greater dependency upon cyber-controlled systems than potential adversary nations. Other nations such as South Korea or Estonia may have greater consumer access to broadband. Others such as the United Arab Emirates may have more Internet-capable mobile devices per capita. But few nations have used computer networks as extensively to control electric power, pipelines, airlines, railroads, distribution of consumer goods, banking, and contractor support of the military.

Second, few nations, and certainly none of our potential adversaries, have more of their essential national systems owned and operated by private enterprise companies. Third, in no other major industrialized and technologically developed nation are those private owners and operators of infrastructure so politically powerful that they can routinely prevent or dilute government regulation of their operations. The American political system of well-financed lobbying and largely unconstrained political campaign contributions has greatly empowered private industry groups, especially when it comes to avoiding meaningful federal regulation.

Fourth, the U.S. military is highly vulnerable to cyber attack. The U.S. military is "netcentric," bringing access to databases and information further down into the operation of every imaginable type of military organization. Along with that access to information systems has come dependence upon them. One small sign of things to come was reported in late 2009. Insurgents in Iraq had used twenty-six-dollar software to monitor the video feeds of U.S. Predator drones through an unencrypted communications link. While not directly threatening to American troops, the discovery raises questions about the Pentagon's beloved new weapon. What if the unencrypted signal could be jammed, thus causing the drone to return home? American forces would be denied one of their most valuable tools and an off-the-shelf program would defeat the product of millions of dollars of research and development. U.S. forces, in addition to being more wired, are also more dependent upon private-sector contractor support than any likely adversary. Even if the U.S. military's own networks were secure and reliable, those of its contractors, who often rely upon the public Internet, may not be.

Those four asymmetries, taken together, tell us that if we and a potential adversary engaged in unlimited cyber warfare, they might do more damage to us than we could do to them. Having some effective limits on what nations actually do with their cyber war knowledge might, given our asymmetrical vulnerabilities, be in the U.S. national interest. Putting that broad theory into practice, however, would require some precise definitions of what kinds of activity might be permitted and what kinds prohibited.

Often arms control negotiations have found difficulty in achieving agreement on something as basic as a definition of what it is that they were seeking to limit. I sat around the table for months with Soviet counterparts trying to define something as simple as "military personnel." For the purposes of discussion in this book, we won't have that kind of delay. Let's take the definition we used in chapter 1 and make it sound more like treaty language:

> Cyber warfare is the unauthorized penetration by, on behalf of, or in support of, a government into another nation's computer or network, or any other activity affecting a computer system, in which the purpose is to add, alter, or falsify data, or cause the disruption of or damage to a computer, or network device, or the objects a computer system controls.

With that definition and the U.S. asymmetrical vulnerabilities in mind, are there successes in other forms of arms control that could be ported into cyberspace, or new ideas unique to the characteristics of cyber war that could form the basis of beneficial arms control? What are the pitfalls of bad arms control to which we should give special attention and caution when thinking about limits on cyber war? How could an international agreement limiting some aspects of cyber war be beneficial to the United States, as well as operationally feasible and adequately verifiable?

SCOPE: ESPIONAGE OR WAR?

Any potential international agreement limiting or controlling cyber war must begin with the scope of the proposal. In other words: What is covered and what is out? The definition of cyber war I used above does not include cyber espionage. Hacking your way in to spy, to collect information, does not add or alter data, nor does it need to damage or disrupt the network or things that the network controls in physical space, if it's done well.

The Russian cyber arms control proposal, however, is sweeping in its scope and would prohibit something that the Russian Federation is doing every day, spying through hacking. The chief public advocate of the Russian proposal, Vladislav Sherstyuk, had a career of managing hackers. As Director of FAPSI, General Sherstyuk was the direct counterpart to the U.S. Director of the National Security Agency. His career background does not *necessarily* mean that General Sherstyuk is now being disingenuous when he advocates an international regime to prohibit what he has directed his agency to do for years. The technical differences between cyber espionage and destructive cyber war are so narrow, perhaps General Sherstyuk thinks that a distinction between the two cannot effectively be made. Or perhaps he has had a change of heart. Perhaps he believes that cyber espionage is something that now puts Russia at a disadvantage. More likely, however, the general, like all who have seen cyber espionage in action, would be very reluctant to give it up.

Cyber espionage is, at one level, vastly easier than traditional espionage. It is hard to exaggerate the difficulty of recruiting a reliable spy and getting such an agent into the right place in an organization so that he or she can copy and exfiltrate a meaningful amount of valuable information. Then there is always the suspicion that the material being provided is falsified and that the spy is a double agent. The best counterintelligence procedure has always been to imagine where the opponent would want to have a spy and then give them one there. The agent passes on low-grade data and then adds some slightly falsified material that makes it useless, or worse.

As I discussed in *Your Government Failed You*, the U.S. is not particularly good at using spies or, as the Americans like to call it, human intelligence (often shortened to HUMINT). The reasons have to do with the difficulty of the task, our reluctance to trust some kinds of people who might make good spies, the reticence of many Americans to become deep-cover agents, and the ability of other nations to detect our attempts at spying. These conditions are deeply seated and cultural, have been true for sixty years or more, and are unlikely to change.

What we are remarkably good at is electronic spying. In fact, our abilities in cyber espionage often make up for our inabilities in the area of HUMINT. Thus, one could argue that forcing the U.S. to give up cyber espionage would significantly reduce our intelligence-collection capability, and that such a ban would possibly put us at a greater disadvantage than it would some other nations.

The idea of limiting cyber espionage requires us to question what is wrong with doing it, to ask what problem is such a ban intended to solve. Although Henry Stimson, Secretary of State under President Herbert Hoover, did stop some espionage on the grounds that "gentlemen do not read each other's mail," most U.S. Presidents have found intelligence gathering essential to their conduct of national security. Knowledge is power. Espionage is about getting knowledge. Nations have been engaging in espionage at least since biblical times. Knowing what another nation's capabilities are and having a view into what they are doing behind closed doors usually contributes to stability. Wild claims about an opponent can lead to tensions and arms races. Spying can sometimes calm such fears, as when in 1960 there was discussion of a "missile gap," that is, that the Soviets' missile inventory greatly exceeded our own. Our early spy satellites ended that concern. Espionage can also sometimes prevent surprises and the need to be ready, on a hair trigger, in constant expectation of certain kinds of surprises. Yet there are some fundamental differences between cyber espionage and traditional spying that we may want to consider.

During the Cold War, the United States and the Soviet Union each spent billions spying on each other. We worked hard, as did the Soviets, to recruit spies within sensitive ministries in order to learn about intentions, capabilities, and weaknesses. Sometimes we succeeded and reaped huge benefits. More often than not, we failed. Those failures sometimes came with damaging consequences.

In the late 1960s, U.S. espionage efforts against North Korea almost led to combat twice. The U.S. Navy electronic espionage ship *Pueblo* was seized, along with its eighty-two crew members, by the North Korean Navy in January 1968. For eleven months, until the crew was released, militaries on the Korean Peninsula were on high alert, fearing a shooting war. Five months after the crew's release, a U.S. Air Force EC-121 electronic espionage aircraft was shot down off the North Korean coast, killing all thirty-one Americans on board (interestingly, on the birthday of North Korean leader Kim Il-sung). The U.S. President, Richard Nixon, considered bombing in response, but with the U.S. Army tied down in Vietnam, he held his fire, lest the incident escalate into a second U.S. war in Asia.

Seven months later, a U.S. Navy submarine was allegedly operating inside the territorial waters of the Soviet Union when the ship collided underwater with a Red Navy submarine. Six years later Seymour Hersh reported, "The American submarine, the USS *Gato*, was on a highly classified reconnaissance mission as part of what the Navy called the Holystone program when she and the Soviet submarine collided fifteen to twenty-five miles off the entrance to the White Sea." According to Peter Sasgen's excellent *Stalking the Bear*, "Operation Holystone was a series of missions carried out during the Cold War [that] encompassed everything from recording the acoustic signatures of individual Soviet submarines to collecting electronic communications to videotaping weapon tests." Both these incidents of spying gone wrong could have brought us into very real and dangerous conflict.

In early 1992, I was an Assistant Secretary of State, and my boss, Secretary of State James A. Baker III, was engaged in delicate negotiations with Russia about arms control and the end of the Cold

War. Baker believed he was succeeding in overcoming the feelings of defeat and paranoia in the leadership circles and the military elites in Moscow. He sought to assuage fears that we would take advantage of the collapse of the Soviet Union. Then, on February 11, the USS *Baton Rouge*, a nuclear submarine, collided not far off the coast from Severomorsk with the Red Banner Fleet's *Kostroma*, a Sierra-class submarine. The Russians, outraged, charged that the U.S. submarine had been collecting intelligence inside the legal limit of their territory.

I recall how furious Baker was as he demanded to know who in the State Department had approved the *Baton Rouge*'s mission and what possible value it could have compared to the damage that could be done by its discovery. Baker urgently embarked on a diplomatic repair mission, promising his embarrassed counterpart, Eduard Shevardnadze, that any future such U.S. operations would be canceled. The USS *Baton Rouge*, badly damaged, made it back to port, where it was, shortly after, struck from the fleet and decommissioned. Those in Moscow who had been preaching that America was hoodwinking them had their proof. The distrust Baker sought to end only grew instead.

As we think of cyber espionage, we should not just think of it as a new intercept method. Cyber espionage is in many ways easier, cheaper, more successful, and has fewer consequences than traditional espionage. That may mean that more countries will spy on each other, and do more of it than they otherwise would.

Prior to cyber espionage, there were physical limits to how much information a spy could steal and, thus, in some areas there were partial constraints on the extent of the damage he could do. The case of the F-35 fighter (mentioned briefly above, in chapter 5) demonstrates how when the quantitative aspect of espionage changes so much with the introduction of the cyber dimension, it does not just add a new technique. Rather, the speed, volume, and global reach

of cyber activities make cyber espionage fundamentally and quali-
tatively different from what has gone before. Let's look at the F-35
incident again to see why.

The F-35 is a fifth-generation fighter plane being developed by
Lockheed Martin. The F-35 is meant to meet the needs of the Navy,
Air Force, and Marines in the twenty-first century for an air-to-
ground striker, replacing the aging fleet of F-16s and F-18s. The
F-35's biggest advancement over the fourth-generation aircraft will
be in its electronic warfare and smart weapons capabilities. With a
smaller payload than its predecessors, the F-35 was designed around
a "one shot, one kill" mode of warfare that depends on advanced
targeting systems. Between the Air Force, Navy, and Marines, the
U.S. military has ordered nearly 2,500 of these planes, at a cost of
over $300 billion. NATO nations have also ordered the aircraft.
The F-35 would provide dominance over any potential adversary for
the next three decades. That dominance could be challenged if our
enemies could find a way to hack it.

In April 2009, someone broke into data storage systems and
downloaded terabytes' worth of information related to the development of the F-35. The information they stole was related to the
design of the aircraft and to its electronics systems, although what
exactly was stolen may never be known because the hackers covered
their tracks by encrypting the stolen information before exporting
it. According to Pentagon officials, the most sensitive information
on the program could not have been accessed because it was allegedly air-gapped from the network. With a high degree of certainty,
these officials believe that the intrusion can be traced back to an
IP address in China and that the signature of the attack implicates
Chinese government involvement. This was not the first time the
F-35 program had been successfully hacked. The theft of the F-35
data started in 2007 and continued through 2009. The reported
theft was "several" terabytes of information. For simplicity's sake,

let's assume it was just one terabyte. So, how much did they steal? The equivalent of ten copies of the *Encyclopaedia Britannica*, all 32 volumes and 44 million words, ten times over.

If a Cold War spy wanted to move that much information out of a secret, classified facility, he would have needed a small moving van and a forklift. He also would have risked getting caught or killed. Robert Hanssen, the FBI employee who spied for the Soviets, and then the Russians, starting in the 1980s, never revealed anywhere near that much material in over two decades. He secreted documents out of FBI headquarters, wrapped them in plastic bags, and left them in dead drops in parks near his home in Virginia. In all, Hanssen's betrayal amounted to no more than a few hundred pages of documents.

Hanssen now spends twenty-three hours a day in solitary confinement in his cell at the supermax prison in Colorado Springs. He is allowed no letters, no visitors, no phone calls, and when addressed by prison guards, is referred to only as "the prisoner" in the third person ("the prisoner will exit his cell"). At least Hanssen escaped with his life. The spies he betrayed were not so lucky. At least three Russians in the employ of the American intelligence community were betrayed by Hanssen and killed by the Russians. A fourth was sent to prison. Spying used to be a dangerous business for the spies. Today it is done remotely.

The spies who stole the information on the F-35 didn't need to wait for a recruit to be promoted to gain access, they didn't have to find someone motivated to betray his country, and no one had to risk getting caught and going to a supermax, or worse. Yet with the information stolen, they may be able to find a weakness in the design or in the systems of the F-35. Perhaps they will be able to see a vulnerability to a new kind of cyber weapon they will use in a future war to eliminate our dominance in the air by dominating cyberspace. That may not even be the worst-case scenario. What if, while the hackers were in our systems, exfiltrating information, they also uploaded a software package? Maybe it was designed to provide a trapdoor for access to the network later, once their original way in was patched. Maybe it was a logic bomb set to take down the Pentagon's network in a future crisis. Moving from espionage to sabotage is just a few clicks of the mouse. Whoever "they" are, they may be in our systems now just to collect information, but that access could allow them to damage or destroy our networks. So, knowing that nations have been in our systems "just to spy" may give the Pentagon and the President a moment of pause in the next crisis.

Banning cyber espionage effectively would present huge challenges. Detecting whether a nation is engaging in cyber espionage may be close to impossible. The ways in which the U.S. and Russia now engage in cyber espionage are usually undetectable. Even if we had means of noticing the most sophisticated forms of network penetration, it could be exceedingly difficult to prove who was on the keyboard at the other end of the fiber, or for whom he was working. If we agreed to a treaty that stopped cyber espionage, U.S. agencies would presumably cease such activity, but it is extremely doubtful that some other nations would.

The ways in which we collect information, including by cyber espionage, may offend some people's sensibilities and may sometimes violate international or national laws, but, with some notable exceptions, U.S. espionage activities are generally necessary and beneficial to U.S. interests. Moreover, the perception that espionage is vital is widespread among U.S. national security experts and legislators. One question I always asked my teams when I was engaged in arms control was, "When it comes time to testify in favor of the ratification of this agreement, how will you explain to the U.S. Senate how you came to agree to this provision, or, since it will likely be me testifying, how the hell do I explain why we agreed to this?" With an agreement to limit espionage, I would not even know where to begin.

And so, when looking at a Russian proposal to ban cyber espionage, one is left wondering why they proposed it and what it says about the overall intent and purpose of their advocacy of a cyber war treaty. The Russian proposal to ban cyber espionage comes from a country with a high degree of skill in such activity, a nation that has regularly orchestrated cyber warfare against other states, has one of the worst records when it comes to international cooperation against cyber crime, and has not signed the one serious international agreement on disruptive cyber activity (the Council of Europe Cyber Crime Convention).

In rejecting the Russian proposal for an international agreement prohibiting cyber espionage, I recognize that cyber espionage does have the potential to be damaging to diplomacy, to be provocative, and possibly even destabilizing. As former NSA Director Ken Minihan said to me, "We are conducting warfare activities without thinking that it is war." That is dangerous, but there may be other ways to address those concerns. Over the course of the Cold War, the CIA and its Soviet counterpart, the KGB, met secretly and developed tacit rules of the road. Neither side went around assassinating the other's agents. Certain things were generally out of bounds. There may be a parallel in cyber espionage. What I recommend is consideration of quiet understandings. Countries need to recognize that cyber espionage can easily be mistaken for preparation of the battlefield and that such actions may be seen to be provocative. Nations should not do things in cyberspace that they would not do in the real world. If you would not put a group of agents in somewhere to extract the information you are hoping to steal on the Net, you probably should not take it electronically. Because there is so little difference between extraction and sabotage, countries should be careful about where they prowl and what they take in cyberspace.

While espionage targeting government systems may have gotten out of hand, America's real crown jewels are not our government secrets, but our intellectual property. U.S. stockholders and taxpayers spend billions of dollars funding research. China steals the results for pennies on the billions and then takes the results to market. The only real economic edge that the U.S. enjoyed, our technological research prowess, is disappearing as a result of cyber espionage. Calling it "industrial espionage" doesn't alter the fact that it is crime. By hacking commercial organizations around the world to steal non-defense data to increase China's profits, the government in Beijing has become a cleptocracy on a global scale. Even if a major cyber war involving the U.S. never happens, Chinese cyber espionage and intellectual property war may swing the balance of power in the world away from America. We need to make protecting this information a much higher priority, and we need to confront China about its activities.

If consequences can be created for certain kinds of destabilizing cyber espionage, countries may more tightly control who does it, why it is done, and where it is done. Most bureaucrats want to avoid scenes in which they have to explain to an outraged Secretary of State, or similar senior official, how the intelligence value of an exposed covert operation was supposed to outweigh the damage done by its discovery. Thus, while I recognize that some cyber espionage may have the potential to be less valuable than the corresponding amount of damage it may cause, I think that risk is best handled by discussions among intelligence organizations and governments bilaterally, privately. An arms control agreement limiting cyber espionage is not clearly in our interest, might be violated regularly by other nations, and would pose significant compliance-enforcement problems.

BANNING CYBER WAR?

Would it be a good idea, then, to agree to an outright ban on cyber war as defined here (that is, excluding cyber espionage)? An outright ban could, theoretically, prohibit the development or possession of cyber war weapons, but there would be no way to enforce or verify such a ban. A ban could also be articulated as a prohibition on the use of cyber weapons against certain targets or on their deployment prior to the outbreak of hostilities, rather than their mere possession or their use in espionage. To judge whether a ban on conducting cyber war would be in our interest, assuming it could be verified, let's look at some hypothetical cases.

Imagine a scenario similar to the Israeli raid on the Syrian nuclear facility with which this book began. Change the scenario slightly so that it is the United States that wants to prevent some rogue state from developing a nuclear weapon and it is the United States that decides it has to bomb the covert site where the nuclear weapon is going to be made. The U.S. might well have the same kind of capability to turn off an adversary's air defense system by employing a cyber weapon. If we had agreed to a ban on the use of cyber weapons, we would face a choice between, on the one hand, violating the agreement, and, on the other hand, sending in U.S. pilots without having done all that we could in advance to protect them. Few civilian or military leaders in this country would want to have to explain that U.S. aircraft were shot down, U.S. pilots taken prisoner or killed, because even though we could have shut off the adversary's air defense system we did not because of an international agreement.

Or imagine a scenario in which the U.S. was already in a limited shooting war with some nation, as we have been in recent history with such nations as Serbia, Iraq, Panama, Haiti, Somalia, and Libya. The U.S. forces might be in a situation where they could substitute a cyber weapon for conventional explosive, kinetic weapons. The cyber weapon might result in lower lethality and do less physical damage, have less long-lasting effects. An outright ban on the use of cyber weapons would force the U.S. to choose, once again, between violating the agreement and doing some unnecessary damage to the adversary.

A simpler scenario would not involve a shooting war or a U.S. preemptive attack, but rather something as routine as a U.S. ship sailing peacefully in international waters. In this scenario, a U.S. destroyer sailing parallel to the North Korean coast would be attacked by a North Korean patrol boat, which fires missiles at the destroyer. The U.S. ship might have a cyber weapon that could be beamed into the guidance system of the incoming missiles, causing them to veer away. If there were an outright ban on the use of cyber weapons, the U.S. might even be prohibited from using them to defend its forces from an unprovoked attack.

The most difficult scenario in which to show restraint would be if cyber weapons were already being used against us. If an adversary tried to shut down a U.S. military network or weapon system by using cyber techiques, it would be tempting to ignore the international agreement and respond in kind.

The two sides of the case for and against a complete ban on the use of cyber war weapons are clear. If we really believe that a ban on cyber weapons is in the U.S. interest, we should be willing to pay some price to maintain the international standard of not using such weapons. We have been in situations in the past where we might have enjoyed some immediate military advantage by using a nuclear weapon or a chemical or biological weapon, but we have always decided that the larger U.S. interest is in maintaining a global consensus against employing such weapons. Nonetheless, because cyber weapons can be less lethal, banning their use in conjunction with kinetic combat may be hard to justify. If shots are already being fired, using cyber weapons might not be destabilizing or escalatory if (and this is a very big if) their use did not expand the scope of the war. The U.S. military will make the case (strongly) that cyber war weapons are a U.S. advantage and that we have to use our technological advantage to compensate for how thinly our forces are spread around the world and how sophisticated the conventional weapons have become that are in the hands of possible opponents.

Balancing our desire for military flexibility with the need to address the fact that cyber war could damage the U.S. significantly, it may be possible to craft international constraints short of a complete ban. An international agreement that banned, under any circumstances, the use of cyber weapons is the most extreme form of a ban. In the previous chapter, we looked briefly at the proposal of a no-first-use agreement, which is a lesser option. A no-first-use agreement could simply be a series of mutual declarations, or it could be a detailed international agreement. The focus could be on keeping cyber attacks from starting wars, not on limiting their use once a conflict has started. We could apply the pledge to all nations, or only to those nations that made a similar declaration or signed an agreement.

Saying we won't be the first ones to use cyber weapons may in fact have more than just diplomatic appeal in the international arena. The existence of the pledge might make it less likely that another nation would initiate cyber weapons use because to do so would violate an international norm that employing cyber weapons crosses a line, is escalatory, and potentially destabilizing. The nation that goes first and violates an agreement has added a degree of international opprobrium to its actions and created in the global community a presumption of misconduct. International support for that nation's underlying position in the conflict might thus be undermined and the potential for international sanctions increased.

A no-first-use declaration could result in reduced flexibility in many of the kinds of cyber scenarios I discussed above. Waiting to respond in kind once we detected that the cyber weapons had been used in a conflict, or used specifically against us, may also create a disadvantage in the cyber war phase of a conflict.

BANNING ATTACKS ON CIVILIANS?

There are less restrictive approaches than banning the use of cyber weapons, or even forswearing first use. One possibility would be to issue a unilateral declaration or to agree to an international protocol placing civilian targets off limits to nation-states' use of cyber weapons. There is ample precedent in the international laws of war for a limited ban on certain weapons or activities, as well as to treaties that call for the protection of civilians caught up in wars.

In World War I, aircraft were used in combat for the first time. They were mainly employed for reconnaissance, machine-gun strafing of troops, and attacking each other in the air, but some aircraft were used to drop explosives on the enemy. This first, small use of aerial bombing opened the possibility of creating larger aircraft in the future to carry more, and bigger, bombs. Within a decade bomber aircraft were being manufactured. One of the earliest science fiction authors, H. G. Wells, vividly portrayed what such bombing aircraft could do to a city in his 1933 novel *The Shape of Things to Come*. By 1936 he and the filmmaker Alexander Korda had adapted the book into a movie, *Things to Come*, which horrified audiences. In 1938 in Amsterdam, an international conference agreed to limits on "New Engines of War." That agreement led, later that year, to a "Convention for the Protection of Civilian Populations against Bombing from the Air."

Unfortunately for Amsterdam, and most major cities in Europe and Asia, that agreement did not stop Germany, Japan, the United States, the United Kingdom, or the Soviet Union from aerial carpet bombing of cities in the war that started one year later. After World

War II, nations tried again and wrote several agreements limiting
how future wars should be conducted. These treaties, negotiated
in Switzerland, became known as the Geneva Conventions. Convention Four covers the "Protection of Civilian Persons in Time of
War." Thirty years later the United Nations sponsored another series
of conventions that protected not only civilians, but also military
personnel against certain kinds of weapons that were thought to
be destabilizing or heinous. The conventions were given the cum-
bersome title "Prohibitions or Restrictions on the Use of Certain
Conventional Weapons . . . Excessively Injurious or Hav[ing] Indiscriminate Effects." Five specific protocols were agreed on, banning
or limiting the use of established weapons such as land mines and
incendiaries, as well as the new application of commercial laser technology to weaponry.

More recently the International Criminal Court agreement,
which entered into force in 2002, banned intentionally targeting
civilians. The United States has withdrawn from the Court treaty
and has gained agreement from many nations that they would not
support the prosecution of U.S. military personnel by the Court.

Either the Geneva convention on "Protection of Civilians" in war
or the UN convention on weapons with "Indiscriminate Effects"
could be expanded to deal with this new kind of warfare. Cyber
weapons used against a nation's infrastructure would inevitably
result in attacking civilian systems. Nothing could be more indiscriminate that attacking such things as a nation's power grid or transportation system. While such broad-based attacks would diminish a
nation's military capacity, some military capabilities will suffer less
than similar civilian infrastructure. The military are more likely to
have backup power systems, stockpiled food, and emergency field
hospitals. A broad-based cyber attack on a nation's infrastructure
could keep the power grid off-line for weeks, pipelines unable to
move oil and gas, trains sidelined, airlines grounded, banks unable

to dispense cash, distribution systems crippled, and hospitals working at severely limited capacity. Civilian populations could well be
left in cold, darkened dwellings with little access to food, money,
medical care, or news about what was happening. Looting and a
crime wave could follow. The number of fatalities would depend
upon the duration and geographic scope of the outages. While such
casualties would, however, be far fewer than those resulting from
an aerial bombing campaign against cities, a sophisticated national
cyber attack would definitely affect civilians, and might even be designed to do so.

Extending existing international agreements to protect civilians against cyber attacks has advantages for the United States.
It allows the U.S. to continue to do what it is good at, cyber war
against military targets, including going first. Sophisticated cyber
weapons may allow the U.S. to continue to have technological superiority in potential military conflicts, even as other nations deploy modern conventional weapons with capabilities that approach
or equal those of American forces. Cyber weapons may also allow
the U.S. to compensate in local or regional situations where the
American forces are outnumbered.

Limiting U.S. cyber attacks to military targets would mean that
we could not disrupt another nation's military as a side effect of a
general attack on a civilian power grid or railroad system. It is likely,
however, that U.S. cyber warriors have the capability to narrowly attack military targets such as command and control grids, air defense
networks, and specific weapons systems. Thus, by respecting a ban
on attacking civilian targets, the U.S. may not lose much or any capability needed that they need to dominate an adversary.

The U.S. is not very good at cyber defense, nor is anybody else;
but the U.S. civilian infrastructure is more vulnerable, and thus
the U.S. stands to suffer more from a broad national cyber attack
than would most other nations. Because the U.S. military relies

on the civilian infrastructure, a ban on cyber attacks on civilian targets would protect the U.S. military, as well as what it would do to avoid inflicting harm on people in general and on the economy.

If the U.S. thought such a limited ban on cyber weapons was in its interest and either proposed it or agreed to it, there are two immediate follow on questions. First, how do you propose to verify it? Let's get to that in a moment. Second, what does it mean with regard to "preparation of the battlefield"? Do we define an attack as including the penetration of a network, or the emplacement of a logic bomb, or is it just the *use* of a logic bomb or other weapon? Specifically, what is it that we would be willing to agree to stop doing?

Earlier, we came to the conclusion that a formal international agreement banning cyber espionage was probably not a good idea for the United States. So, we would not ban the penetration of networks to collect intelligence, and there is probably intelligence information that one could glean from hacking into a railroad's control system. But what real intelligence value would there be to hacking into an electric grid's controls? Hacking into an electric grid's controls and leaving a trapdoor to facilitate easy return can have only one purpose: preparation for an attack. Leaving behind a logic bomb is even more obviously an act of cyber war.

Theoretically, you could write a ban on cyber war attacks on civilian infrastructure that would not explicitly prohibit placing trapdoors or logic bombs, but would rather just ban any act that actually causes a disruption. This narrow ban would allow the U.S. to be in position to retaliate quickly against another country's civilian infrastructure if it attacked ours. Without preplacement of cyber weapons, it might be difficult and time-consuming to attack networks. But by allowing countries to go around lacing one another's networks with logic bombs, we would be missing the chief value of a ban on cyber attacks on civilian infrastructure.

The main reason for a ban on cyber war against civilian infrastructures is to defuse the current (silent but dangerous) situation in which nations are but a few keystrokes away from launching crippling attacks that could quickly escalate into a large-scale cyber war, or even a shooting war. The logic bombs in our electric grid, placed there in all likelihood by the Chinese military, and similar weapons the U.S. may have or may be about to place in other nations' networks, are as destabilizing as if secret agents had strapped explosives to transmission towers, transformers, and generators. The cyber weapons are harder to detect; and, with a few quick keystrokes from the other side of the globe, one disgruntled or rogue cyber warrior might be able to let slip the dogs of war with escalating results, the limits of which we cannot know.

Although we can imagine situations in which the U.S. might wish it had already put logic bombs in some nation's civilian networks, the risks of allowing nations to continue this practice would seem to far outweigh the value of preserving for ourselves that one option to attack. Thus, as part of a ban on attacking civilian infrastructure with cyber weapons, we should probably agree that the prohibition include the penetration of civilian infrastructure networks for the purpose of placing logic bombs, and even the emplacement of trapdoors on networks that control systems such as electric power grids.

BEGINNING WITH THE BANKS?

Even an agreement limited to protecting civilian infrastructure may pose problems. Some nations, like Russia, might contend that a U.S. willingness to accept such an agreement confirms their point that cyber weapons are dangerous. They could hold out for a complete ban. Negotiating a verification arrangement for even a civilian-protection protocol could, as we will discuss shortly, open a Pandora's box of complications. Therefore, the U.S. may want to consider an even more limited scope for an initial international agreement on cyber weapons. One option might be an accord designed to preclude cyber attacks on the international financial system. Every major nation has a stake in the reliability of the data that underpin international bank clearinghouses, their major member banks, and the major stock and commodity trading exchanges. With few exceptions, such as the impoverished rogue state of North Korea, to launch an attack on an element of the international financial system would likely be self-defeating. The damage to the system could directly hurt the attacker, and certainly the financial retaliation that would result from the identification of an attacking nation could cripple a nation's economy.

Because of the interlocking nature of major global financial institutions, including individual banks, even a cyber attack on one nation's financial infrastructure could have a fast-moving ripple effect, undermining confidence globally. And, as one Wall Street CEO told me, "It is confidence in the data, not the gold bullion in the basement of the New York Fed, that makes the world financial markets work."

The belief that cyber attacks on banks could unravel the entire global financial system has prevented successive U.S. administrations from approving proposals to hack into banks and steal funds from terrorists and dictators, including Saddam Hussein. As Admiral McConnell has noted, "What happens if someone who is not deterred attacks a large bank in New York and contaminates or destroys the data? Suddenly there is a level of uncertainty and loss of confidence. Without confidence that transactions are safe and will reconcile, financial transactions will stop." Thus, since we seem to have a self-imposed ban anyway, it would probably be in the interest of the United States to propose or participate in an international agreement to forswear cyber attacks targeted on financial institutions. (Such an agreement need not prohibit cyber espionage. There might be intelligence value from observing financial transactions in banks, such as identifying the money of terrorists. The U.S. may already be doing just that. It apparently came as a shock to European financial institutions in 2006 that the U.S., seeking to track terrorist funds, may have been covertly monitoring the international financial transactions of the SWIFT bank-clearing system.)

INSPECTORS IN CYBERSPACE

The value of international agreements to ban certain kinds of cyber warfare activities, or pledges not to engage in such attacks first, may depend in part upon whether violations can be detected and whether blame can be assigned. Traditional arms control verification is very different from anything that would work in cyberspace. To verify compliance with numerical limits on submarines or missile silos, nations had only to fly their space-based surveillance platforms overhead and take photographs. It's hard to hide a submarine-building shipyard or a missile base. For smaller objects, such as armored combat vehicles, inspection teams were permitted into military bases to conduct inventories. To ensure no improper activity at nuclear reactors, the International Atomic Energy Agency's inspectors install surveillance cameras and place seals and identification tags on nuclear material. International teams sample chemicals at corporations' chemical plants, looking for signs of covert chemical weapons production. To monitor for nuclear weapons tests, an international network of seismic sensors has been netted together, with nations sharing the data they detect.

Only that seismic network, and perhaps the IAEA teams, offers any useful precedent for cyber arms control verification. You cannot detect or count cyber weapons from space, or even by driving around an army base. No nation is likely to agree to having international teams of inspectors plowing through what programs are on computer networks designed to protect classified information. Even if in some parallel universe, nations did permit such intrusive inspection of their military or intelligence computer networks, a nation could hide its cyber weapons on thumb drives or CDs anywhere in the country. A ban on development, possession, or testing of cyber weapons on a closed network (such as the National Cyber Range being developed by Johns Hopkins University and Lockheed Martin) is not something that could be verified.

The actual use of cyber weapons, however, may be more clearcut. The effects of an attack can often be easily discerned. Computer forensic teams can generally determine what attack techniques were used, even if they may not be able to determine how the penetration into the network occurred. The attribution problem would persist, however, even in the case of an attack that has already taken place. Trace-back techniques and ISP records may indicate that a particular nation is involved, but they would not usually be able to prove a government's guilt with high confidence. A nation, perhaps the U.S., could easily be framed. Cyber attacks against Georgia, probably orchestrated by Russia, came from a botnet control computer in Brooklyn.

Even if a nation admitted that an attack came from computers on its territory, the government could claim the attacks were from anonymous citizens. This is precisely the claim that the Russian government did make in the case of the cyber attacks on Estonia and Georgia. It is exactly what the Chinese government claimed when U.S. networks were hit from China in 2001, following the alleged penetration of Chinese airspace by a U.S. electronic spy plane. It may even be true that the hackers would turn out to be people without government jobs or offices, although they may have been encouraged and enabled by their governments.

One way to address the attribution problem is to shift the burden from the investigator and accuser to the nation in which the attack software was launched. This same burden shifting has been used in dealing with international crime and with terrorism. In December 1999, Michael Sheehan, then the U.S. ambassador for counterterrorism, had the job of delivering a simple message to the Taliban. Sheehan was instructed to make it clear to the Taliban that they would be held responsible for any attack perpetrated by al Qaeda against the United States or its allies. Late at night, Sheehan delivered the message through an interpreter by telephone to a representative of the Taliban leader Mullah Omar. To drive home the point, Sheehan used a simple analogy: "If you have an arsonist in your basement; and every night he goes out and burns down a neighbor's house, and you know this is going on, then you can't claim you aren't responsible." Mullah Omar did not evict the arsonist in his basement, indeed he continued to harbor bin Laden and his al Qaeda followers even after 9/11. Now it is Mullah Omar who is huddling in a basement somewhere, hunted by NATO, U.S., and Afghan armies.

The notion contained in the "arsonist principle" is one that can be applied to cyber war. While we talk about cyberspace as an abstract fifth dimension, it is made up of physical components. These physical components, from the high-speed fiber-optic trunks, to every router, server, and "telecom hotel," are all in sovereign nations, except perhaps for the undersea cables and the space-based relays. Even they are owned by countries or companies that have real-world physical addresses. Some people like to contend that there is a "sovereignty problem" on the Internet, that because no one owns cyberspace in its entirety, no one has any responsiblility for its integrity or security. The arsonist principle, articulated in an international agreement as National Cyberspace Accountability, would make each person, company, ISP, and country responsible for the security of their piece of cyberspace.

At a minimum, countries like Russia could no longer claim that they have no control over so-called patriotic hacktivists. An international agreement could hold host governments responsible either for stopping these hackers from participating in illegal international activities, or at least requiring nations to make their best effort to do so. In addition to their own police activities, a nation that is party to an international agreement might have an *obligation to assist*. Such an obligation could require them to respond quickly to inquiries in international investigations, seize and preserve server or router records, host and facilitate international investigators, produce their citizens for questioning, and prosecute citizens for specified crimes.

The existing 2001 Council of Europe Convention on Cyber Crime already incorporates many of these obligations to assist. The United States is a party to the convention. Our sovereignty is not being infringed upon by some supranational Olde Europa bureaucracy. Rather, by signing the convention, the U.S. is promising to pass any new legislation necessary to provide the U.S. government with the authority to do the things necessary to meet the obligations in the agreement.

Going beyond the current cyber crime convention, however, a cyber *war* convention could make nations responsible for ensuring that their ISPs deny service to individuals and devices participating in attacks and report them to authorities. Such a provision would mean that ISPs would have to be able to detect and "black-hole" major worms, botnets, DDOS attacks, and other obvious malicious activity. (Some of this process of identifying malware is something far less difficult than deep-packet inspection and can be done largely by something called "flow analysis," which really means nothing more than watching how much traffic is moving on the network and looking for unusual spikes or patterns.) If a nation did not successfully compel an ISP into compliance, the international agreement could establish a procedure that transferred responsibility to other nations. An ISP could be internationally black-listed. All participating nations would then be required to refuse traffic going to or from that ISP until it complied and stopped the botnets or other obvious malware.

Such an international agreement would deal with a portion of the attribution problem, by shifting responsibility. Even if the attacker could not be identified, at least there would be someone who could be held responsible for stopping the attack and investigating who the attacker was. Such an obligation would not require most nations to add new cyber forensics units. Nations like China and Russia have the ability now to identify and move quickly against hackers. As Jim Lewis of the Center for Strategic and International Studies has said, "If a hacker in St. Petersburg tried to break into the Kremlin system, that hacker could count the remaining hours of his life on one hand." You can be sure that the same is true for anyone in China trying to hack the People's Liberation Army network. If China and Russia signed a cyber war agreement with obligations like the ones suggested here, those governments could no longer blame their citizens for DDOS attacks on other nations and then stand back and do nothing. Failure to act promptly against citizen hackers would result in the nation itself being held in violation of the agreement and, more important, in other nations disconnecting all traffic from the offending ISPs. Nations could black-hole such rogue traffic from other countries now, but in the absence of a legal framework, they are reluctant to do so. An agreement would not only permit nations from blocking such traffic, it would require them to do so.

A National Cyberspace Accountability provision and its corollary Obligation to Assist would not completely solve the attribution problem. The Russian botnet attack could still come from Brooklyn. The Taiwanese hacker sitting in the San Francisco cyber café could still attack a Chinese government website. But under such an agreement the U.S. would have to stop the botnet and actively investigate the hacker. In the case of a hypothetical Taiwanese agent hacking into Chinese networks in violation of an international agreement, the U.S. government, when notified by China of such activity, would have to task the FBI or Secret Service to help the Chinese police track down the culprit in San Francisco. If he was found, he could be tried in a U.S. court for violation of U.S. law.

Of course, nations may say that they are looking for hackers and not be. They may try culprits and find them not guilty. When notified of a botnet originating on an ISP in their country, nations may take their sweet time doing something about it. To judge whether a nation is actively complying or is just being passive-aggressive, it may be useful if a cyber war agreement created an "International Cyber Forensics and Compliance Staff." The staff of experts could make reports to member states on whether or not a nation is acting in the spirit of the agreement. There could be international inspection teams, similar to those under the nuclear nonproliferation agreement, the chemical weapons ban, and the European security and cooperation agreement. Such teams could be invited in by signatory nations to assist in verifying that a cyber war attack had occurred in violation of the agreement. They could help determine what nation had actually launched the attack. The international staff might also, with the voluntary cooperation of member states, place traffic-flow monitoring equipment at key nodes leading into a nation's networks to help detect and identify the origin of attacks.

The international staff might also run a center that nations could contact whenever they believed they were coming under a cyber war attack. Imagine that an Israeli network is hit with a botnet DDOS attack from an ISP in Alexandria, Egypt, at three in the morning, Tel Aviv time. Israel, like all signatory countries in our hypothetical agreement, would have a national cyber security liaison office constantly staffed. The Israeli center would call the international center, say, in Tallinn, and report that a cyber attack was originating from a certain ISP in Egypt. The international center would then call the Egyptian national center in Cairo and request that they immediately investigate whether there is a botnet operating on that ISP in Alexandria. The international staff would time how long it took Egypt to comply and shut down the attack. Perhaps the international staff would be able to look at traffic-flow monitors on gateways coming out of Egypt and see the botnet spike. Egypt would be required to respond with a report on its investigation of the attack. If the incident warranted it, the international staff might ask to send a team of investigators to assist or observe the Egyptian authorities. The international staff could file a report, with conclusions and recommendations, to member states on the incident.

Nations that were found to be scofflaws could be subject to a range of sanctions. In addition to having traffic to and from offending ISPs denied by ISPs in other member states, the offending nation could have its hands slapped by the international organization. For more drastic action, nations could deny visas to officials from the offending nation, limit exports of new IT equipment to the nation, limit the overall amount of cyber traffic to and from the nation, or disconnect the nation altogether from international cyber space for a period of time.

These verification and compliance provisions in a cyber war agreement would not totally solve the attribution problem. They would not prevent a nation from spoofing the source of an attack or framing another state. They would, however, make it more difficult for some kinds of cyber war attacks, while establishing norms of international behavior, providing international legal cover for nations to assist, and creating an international community of cooperating experts in fighting cyber war. It is also important to remember that the capability to conduct attacks that amount to cyber war currently requires a state-level effort, and only a handful of states have advanced capabilities. The list of potential attackers is small. Attribution is a major problem for cyber crime, but for warfare, technical forensics and real-world intelligence can narrow down the list of suspects fairly quickly.

What emerges from this discussion of cyber arms control are five broad conclusions. First, unlike other forms of arms control that destroy weapons, cyber arms control cannot eliminate capability. It can only prohibit acts. Thus, a nation could move from a state of compliance to a gross violation in seconds and without warning.

Second, broad definitions of cyber warfare, such as those that include espionage, are not verifiable and are not in our interest as a nation. Nonetheless, national intelligence services and national governments should initiate channels for discussions so that intelligence activities do not get out of hand, or become misconstrued as showing hostile intentions.

Third, international agreements that prohibit certain acts, such as cyber attacks on civilian infrastructure, are in our interest. Because such attacks could still take place, such agreements would not in any way diminish the need to take defensive steps to protect that infrastructure.

Fourth, high-confidence verification of compliance with a cyber war limitation agreement will not be possible. We may be able to verify a violation, but attribution of the attack will be difficult and could be subject to intentionally misleading activity. Nonetheless, there are measures that can contribute to an international norm against cyber attacks on civilians, namely, an expert international staff, national governmental responsibility for the prevention of violations originating within a nation's borders, and an obligation to assist in stopping and investigating attacks.

Finally, limits on cyber war attacks against civilian infrastructure would probably mean that we and other states would have to cease any activity in which we may be engaged with logic bombs, and perhaps trapdoors, in other nations' civilian infrastructure networks.

Lacing infrastructure with trapdoors and logic bombs, although little noticed or discussed by the media and the general population, is dangerously provocative. They are alluring because they offer some of the results of war, but without soldiers or death. But they also signal hostile intent far more than any weapon that stays in a nation's inventory. They could be utilized easily and quickly, without proper authorization, or without a full appreciation for what kind of spiral of escalation they might cause. Although a war might start in cyberspace and be conducted without soldiers or bloodshed, it would be highly unlikely to stay that way for long. By lacing on another's infrastructure networks with cyber weapons, nations have made starting a war far too easy.

CHAPTER EIGHT

THE AGENDA

Invisibly, military units from over a score of nations are moving into a new battlespace. Because the units are unseen, parliaments and publics have not noticed the movement of these forces. Because their first skirmishes have been isolated and involved only simple weapons, few have thought that cyber warriors could do more. Because most of the major military powers are also one another's trading partners, commentators cannot envision the circumstances that could turn their relations to hostility. Because the United States has been at war in one nation for seven years and in another for nine, is struggling with its worst-ever recession, and is diverted by partisanship, the "bandwidth" of its policy elites is already consumed. Thus, with attention diverted elsewhere, we may be laying the groundwork for cyber war.

There may be parallels in the early years of the last century.

Barbara Tuchman in *The Proud Tower* describes a world similarly diverted from the realization that its various militaries were preparing devastating forces without contemplating the horrific consequences of their use. Then, as she describes in the sequel, *The Guns of August*, a spark caused those forces to be activated. Von Schlieffen's elaborate military use of Germany's massive new freight rail network literally set wheels in motion that could not be stopped. The military use of the new chemical industry added an element of destructiveness. The use of chemical weapons did far more damage than anyone had anticipated. Today our military is developing elaborate plans for a new kind of war, once again using a technology originally designed for commercial use. As in the period one hundred years ago, those plans have received little public scrutiny.

There have been few times in our history when the American academic community, the media, and the Congress have focused on a potential problem and together cast so much light on an issue that controls were put in place that averted calamity. The issue of strategic nuclear war, referenced much in this book, is the clearest example. A new technology had burst upon the world and the U.S. military had seen in it a way to achieve military dominance and, through that, peace. At airbases with the signs "Peace Is Our Profession," the plans called for early and massive use of nuclear weapons in a war, against cities and civilian targets. Not until the research community focused a public klieg light on those plans and the larger issue of how to fight nuclear war, were rational controls and plans developed and adopted.

Today at U.S. Cyber Command, and at its related agencies, some of our nation's most intelligent, patriotic, and undercompensated government employees, military and civilian, are putting plans and capabilities in place to achieve "dominance in cyberspace" to maintain this country's security and preserve the peace. In other nations, cyber war units are also preparing. As part of that preparation, cyber warriors are placing trapdoors in civilian networks, placing logic bombs in electric power grids, and seeding infrastructure for destruction. They believe that their new form of warfare is an advance, not just because of its use of the latest technology, but because it does not involve explosives and direct lethality. Like the Predator pilots who sit in the United States, killing Taliban in Pakistan by remote control, they could subconsciously think that because they live in a peaceful suburban environment, the effects of their destruction on the other side of the world may somehow be clean and neat, unlike "real war."

When in a period of rising tensions, in some future crisis now unforeseen, a cyber warrior of some nation is ordered to "send a message" to the potential adversary by using one of the logic bombs already in place, will it forestall or will it trigger a broader shooting war? Perhaps because the opponent is misled about who started the war, other nations will be drawn in. Possibly, the cyber warrior in one of the score of nations with capability will act without authority, initiating a conflict. Alternatively, it may be a hacker who uses a cyber weapon for destruction rather than crime, or discovers and sets off a logic bomb left behind by someone else. The cyber war that ensues could be incredibly rapid and global.

When an American President sends U.S. forces to bomb a rogue state's nuclear weapons factory or terrorist camp, that nation may not be able to respond against our impressive conventional military forces. And yet, for a small investment in a cyber war capability, it may respond by destroying the international financial system, in which it has very little stake. The asymmetry of what it costs to counter our conventional military versus the minimal investment required for a cyber war capability will tempt other nations, and perhaps criminal cartels and terrorist groups as well.

Because the U.S. invented the Internet and has perhaps led in cyber espionage and the creation of cyber war tools, it may have developed an implicit arrogance, causing us to assume that no one could humble America in a cyber war. Our cyber warriors and, to the extent that they think of cyber war, our national security leaders in general, may take comfort in the fact that we could perhaps see a cyber attack coming. They may think that we could block some of it, and they may believe we could respond in kind, and then some. The reality is that a major cyber attack from another nation is likely to originate in the U.S., so we will *not* be able to see it coming and block it with the systems we have now or those that are planned. Yes, we may be able to respond in kind, but our nation will still be devastated by a massive cyber attack on civilian infrastructure that smacks down power grids for weeks, halts trains, grounds aircraft, explodes pipelines, and sets fires to refineries.

The reality may also be that when the U.S. President wants to retaliate further, he will be the one who will have to escalate. He will be the one who will have to cross the cyber/kinetic boundary. And he may find, when he does, that even our conventional forces are cyber dependent. The U.S. military's reliance upon cyber systems exceeds the extensive dependence of the commercial infrastructure. The contractors required for America to fight a war may be immobilized by cyber attack. The allegedly hermetically sealed computer networks upon which the Department of Defense relies may prove porous and unavailable. Highly advanced technology in the conventional weapons and systems that give U.S. forces dominance (for example, the F-35 fighter and the Global Positioning System) may suddenly not work. We are not the only nation that can install a logic bomb.

With a nation in the dark, shivering in the cold, unable to get food at the market or cash at the ATM, with parts of our military suddenly impotent, and with the regional flashpoint that started it all going badly, what will the Commander-in-Chief do? Perhaps he will appoint a commission to investigate what went wrong. That commission will read the work of another commission, one appointed by Bill Clinton in 1996, and be astonished to learn that this disaster was foreseen back then. They will note the advice of a non-government commission written in 2008 advising the next President to take cyber war seriously. They may, if they are diligent, find a National Academy of Sciences study on Offensive Information Warfare from 2009 that warned that cyber war policy was "ill-formed, undeveloped, and highly uncertain."

The post-disaster commission, a special committee of the Congress, or the next President would likely recommend a plan so that "this sort of thing can never happen again." Since we know now what has been recommended already, what hasn't worked, and why, perhaps we should not wait for a disaster to embark on a plan to deal with cyber war. If we strip away the luxuries and the things that would be nice to have, there are six simple steps that we need to take simultaneously and now to avert a cyber war disaster.

1. THINKING ABOUT THE UNSEEABLE

First, we must initiate a broad public dialogue about cyber war. A student looking to choose a graduate school asked me recently to recommend a university where she could take courses on cyber war. We scoured course catalogues and found none at any of the major security-policy schools, such as Harvard's Kennedy School, Princeton's Woodrow Wilson School, or Texas's Lyndon Johnson School. She asked what books she should read and we found some interesting titles, but few that really delved into the policy and technology of cyber war. Many that seemed promising turned out to use the phrase "information war" to mean psychological warfare or public diplomacy.

Perhaps there are few books on cyber war because so much of

the subject matter is secret. Maybe there should be public discussion precisely because so much of the work has been stamped secret. In the 1950s and 1960s, people like Herman Kahn, Bill Kaufmann, and Albert Wohlstetter were told that nuclear war was something that could not really be discussed publicly. One of Kahn's responses was a book called *Thinking About the Unthinkable* (1962), which contributed to a robust public dialogue about the moral, ethical, and strategic dimensions of nuclear war. Open research and writing done at MIT, Harvard, Princeton, Chicago, and Stanford also contributed. Bill Kaufmann's classes at MIT, Harvard, and the Brookings Institution taught two generations how to think about nuclear strategy and how to ask analytical questions, so that they could think on their own. Today at Harvard and MIT, the aptly named Project Minerva, an open research program on cyber war funded by the Defense Department, has begun. (I am reminded of Hegel's dictum that "the owl of Minerva always flies at dusk," meaning that wisdom comes too late.)

The mainstream media's treatment of cyber war has improved. Reporters at the *Wall Street Journal* and the *New York Times* have written on it since 2008. Public television's highly respected *Frontline* series did an hourlong examination in 2003, *Cyber War*. Television has focused much more on identity theft by cyber criminals because so many readers and viewers have already been victimized by cyber crime. Movies, however, have been filled with cyber war. In *Live Free or Die Hard*, a former government cyber security official who wasn't listened to (whom the *New York Times* reviewer said was reminiscent of me. Nonsense!) cripples national systems. In *Eagle Eye*, hacking causes high-tension lines to melt and general havoc to erupt. In *The Italian Job*, the hacking is limited to traffic lights, but in *Ocean's Eleven* there is a power blackout in Las Vegas. There are so many more that much of the moviegoing public has little trouble understanding what cyber war can do. High-level policy officials apparently seldom make it to the movies. Or maybe they think it's all just fantasy. To make them understand that such scenarios can really happen, we need an exercise program to drive home the point. General Ken Minihan has been promoting the idea of an Eligible Receiver–type war game for the private sector. "We could scare the pants off them, the way we did for the President in '97."

Congress, surprisingly, has held numerous hearings on cyber security and has tasked its Government Accountability Office to investigate. One GAO report asked whether the warnings that hackers could attack a power grid were true. GAO investigated one of the few power grids owned and operated by the federal government, the Tennessee Valley Authority's system. GAO reported back in 2008 that there were significant cyber security vulnerabilities on the TVA grid that left it open to attack. On cyber war, however, as distinct from cyber security in general, Congress has done little in the way of oversight, hearings, or legislation.

Congress is a federation of fiefdoms, subject to the vicissitudes of constant fund raising and the lobbying of those who have donated the funds. That situation has two adverse consequences with regard to congressional involvement in cyber war oversight. First, everyone wants his or her own fiefdom. Congress has resisted any suggestion, such as was made by Senator Bob Bennett (Republican of Utah), that there be one committee authorized to examine cyber security. As a result there are approximately twenty-eight committees and subcommittees involved in the issue and none with jurisdiction to think holistically. Second, Congress "eschews regulation" and spits it out. The influential donors from the information technology, electric power, pipeline, and telecommunications industries have made the idea of serious cyber security regulations as remote as public financing of congressional campaigns or meaningful limits on campaign contributions.

The dialogue we need will require meaningful academic research and teaching, a shelf of new books, in-depth journalism, and serious congressional oversight.

2. THE DEFENSIVE TRIAD

The next item on the agenda to prevent cyber war is the creation of the Defensive Triad. As proposed earlier in this book, the Triad stops malware on the Internet at the backbone ISPs, hardens the controls of the electric grid, and increases the security of the Defense Department's networks and the integrity of its weapons. Much of the work in DoD has already begun as a result of President Bush's decision in his last year in office. The Defensive Triad is not an attempt, as my National Strategy for Cybersecurity was, to defend everything. The Triad is, however, designed to defend enough so as to cause another nation to think twice before launching a cyber war against us. A potential attacker needs to believe that much of his attack will fail and that its greatest effect will be retaliation of various sorts. Without the Defensive Triad, the U.S. should itself be deterred from acting in any way (not just in cyber war) that could provoke someone into a cyber war attack on America. Today we are so vulnerable to a devastating cyber war attack that U.S. leaders should walk cautiously.

We cannot build two of the three prongs of the Defensive Triad (the defense of the Tier 1 ISPs and of the electric power) without additional regulation. The argument I have made in the past about homeland security in general is that without using regulation the federal government is trying to achieve security with one of its arms tied behind its back. There was an era when federal regulations were overly intrusive and ineffective, but that is not inherent in the idea of the government asking industries to avoid doing some things and defining desired end states. At the Black Hat conference in 2009 (discussed earlier), the cyber security expert and author Bruce Schneier made the same point, arguing that "smart regulation" that specifies the goal and does not dictate the path is needed to improve cyber security.

Our cyber war agenda must include regulation that requires the Tier 1 ISPs to engage in deep-packet inspection for malware and to do so with the highest standards of privacy protection and oversight. The ISPs must be given the legal protection necessary so that they do not have to fear being sued for stopping viruses, worms, DDOS attacks, phishing, and other forms of malware. Indeed, they must be required to do so by new regulations.

In order for the Department of Homeland Security to fulfill its role in the Defensive Triad, we must create a reliable and highly qualified component, perhaps a Cyber Defense Administration. The Cyber Defense Administration should be responsible for overseeing the deep-packet inspection system that the ISPs will run. It should also be responsible for monitoring the health of the Internet in real time, take over responsibility for regulating cyber security of the power sector from the Federal Energy Regulatory Commission (FERC), and provide a focal point for law enforcement activities related to cyber crime. The Cyber Defense Administration's most important role, however, would be to manage the defense of both the dot-gov domain and critical infrastructure during an attack.

The administration could provide the ISPs with known signatures of malware in real time, in addition to being a vehicle for the ISPs sharing what they themselves discover. The existing National Communications System, a four-decade-old office that worked on telephone availability in emergencies, and which was recently merged into the new National Cybersecurity and Communications Integration Center (NCCIC, but pronounced "en-kick"), could provide the ISPs with an out-of-band communications system that could pass these malware signatures. The Cyber Defense Administration could draw on the expertise of the Pentagon and intelligence agencies, but the National Security Agency must not be given the mission of protecting domestic U.S. cyber networks. As uniquely skilled as NSA's experts are, they and their agency suffer from a public distrust exacerbated by the warrantless wiretapping ordered by Bush and Cheney.

Beyond regulating the ISPs, the other area of regulation needed is the electric power grid. The only way to secure the grid is to require encryption of commands to the devices running the system, along with authentication of the sender, and a series of completely out-of-band channels that are not connected to the companies' intranets or the public Internet. The FERC has not required that, but it did finally issue some regulations in 2008. It has not yet started to enforce them. When it does, do not expect much. That commission completely lacks the skills and personnel needed to ensure that electric power companies disconnect their controls from any pathway that a hacker could use. The mission of auditing the electric companies' compliance should also be given to the Cyber Defense Administration, where the expertise could be built and where the overly chummy relationship with the industry exhibited by the FERC would not get in the way of security.

The Cyber Defense Administration should also assume the cyber security responsibility for the myriad civilian federal departments and agencies, all of which are now forced to try to do cyber security on their networks. Also, consolidating in the proposed Cyber Defense Administration what is now done on cyber security by the Office of Management and Budget and the General Services Administration would increase the probability of achieving a center of excellence that could manage security on the government's own civilian (not Defense) networks.

3. CYBER CRIME

Because cyber criminals can become rental cyber warriors, we need as the third agenda item to reduce the level of cyber criminality that is plaguing the Internet. Cyber criminals have begun to penetrate the supply chains for both computer hardware and software manufacturers to inject malicious code. Instead of just using widely available hacking tools, cyber criminals are now starting to write their own specially designed code to beat security systems, as was the case in the theft of millions of credit card numbers from T.J. Maxx in 2003. These trends point to the growing sophistication of cyber criminals, and may indicate that the criminal threat could grow to become as sophisticated as the state-level threat. That suggests we need to increase our efforts to combat cyber crime.

Today both the FBI and the Secret Service investigate cyber crime, with help from Customs (now called Immigration and Customs Enforcement, or ICE) and the Federal Trade Commission. Yet companies and citizens across the country complain that their reports of cyber crime go unanswered. The Justice Department's ninety independent prosecutors scattered around the nation often ignore cyber crime because individual cyber thefts usually fall below the $100,000 minimum necessary for a federal case to be authorized. The U.S. attorneys are also often computer illiterate and do not want to investigate a crime where the culprit is in some other city or, worse yet, another country.

The President could assign the FBI and Secret Service agents who cover cyber crime to the proposed Cyber Defense Administration, along with attorneys to prepare cases for the Justice Department. A single national investigatory center within the Cyber Defense Administration, coordinating the work of regional teams, could develop the expertise, detect patterns, and engage in the international liaison needed to increase the probability of arrest to the point where it might begin to be a deterrent. Today law enforcement in the U.S. does not begin to deter the world's cyber criminals. Today cyber crime does pay. To make it stop paying, the U.S. would need to make a substantially greater investment in federal law enforcement agencies' cyber crime capability. We will also have to do something about cyber crime sanctuaries.

In the late 1990s, international criminal cartels were laundering hundreds of billions of dollars through "banks" in a variety of mini-nations, usually island states, as well as several larger sanctuary nations. The major financial powers got together, agreed on a model law criminalizing money laundering, and told the sanctuary states to pass the law and enforce it. If they didn't, the countries were told that the major international financial nations would all stop clearing their local currencies and halt financial transactions with their banks. I had the pleasure of conveying that message to the Prime Minister of the Bahamas, where the law was promptly passed. Money laundering did not disappear, but it got a lot harder because there were fewer reliable sanctuaries. The signatories of the Council of Europe Convention on Cyber Crime should do the same kind of thing to cyber crime sanctuaries. Together they need to tell Russia, Belarus, and the other scofflaws that they either have to start enforcing laws against cyber crime or there will be consequences. One of the consequences would be to limit and inspect all Internet traffic entering nations from the scofflaw sanctuaries. It's worth a try.

4. CWLT

The fourth component of the agenda to address cyber war should be the equivalent of the Strategic Arms Limitation Treaty (SALT) for cyber war, a Cyber War Limitation Treaty, or CWLT (pronounced "*see*-walt"). The U.S. should coordinate the proposal with its key allies in advance of suggesting it at the United Nations. As the name implies, it should limit cyber war, not seek some global ban on hacking or intelligence gathering. SALT and its follow-on Strategic Arms Reduction Treaty (START) not only accepted intelligence collection as an inevitability, they relied upon it and called for "noninterference" with it. Those treaties explicitly protected what they called "national technical means."

When arms control worked well, it had begun somewhat modestly and then expanded its scope in subsequent agreements as confidence and experience had grown. CWLT should begin by doing the following in an initial agreement:

- establish a Cyber Risk Reduction Center to exchange information and provide nations with assistance;
- create as international-law concepts the *obligation to assist* and *national cyber accountability*, as discussed earlier;
- impose a ban on *first-use* cyber attacks against civilian infrastructure, a ban that would be lifted when (a) the two nations were in a shooting war, or (b) the defending nation had been attacked by the other nation with cyber weapons;
- prohibit the preparation of the battlefield in peacetime by the emplacement of trapdoors or logic bombs on civilian infrastructure, including electric power grids, railroads, and so on; and
- prohibit altering data or damaging networks of financial institutions at any time, including the preparation to do so by the emplacement of logic bombs.

Later, after experience with CWLT One, we could examine whether to expand its scope. We should begin with a no-first-use ban on cyber attacks against civilian targets, rather than an outright ban, because nations should not be disingenuous when they sign obligations. Nations that are engaged in a shooting war or have been the victims of cyber attack will probably employ cyber weapons. Moreover, we do not want to force nations that have been the victim of cyber attack to retaliate with kinetic weapons because of a ban on cyber attacks. The proposal does not preclude initial cyber attacks on military targets. Nor does it rule out preparation of the battlefield against military targets, because proposals to do so raise complex trade-offs and would overburden CWLT One. Nonetheless, lacing each other's military with logic bombs is destabilizing and we should say publicly that if we discover it happening to us we would consider it as a demonstration of hostile intent.

Non-state actors will be a problem for cyber arms control, but CWLT should shift the burden of stopping them to the states party to the convention. Nations would be required to rigorously monitor for hacking originating in their country and to prevent hacking activity from inside their territory. They would be required to act promptly to stop such activity when notified of it by other nations through an international Cyber Threat Reduction Center. That Center would be created by the treaty, paid for by signatories, and be staffed at all times by network and cyber security experts. The Center could also dispatch computer forensics teams to assist in investigations and to determine whether nations are actively and assiduously investigating reported violations. The treaty would include a concept of *national cyber accountability*, making it a treaty violation if a nation did not stop a threat when notified by the Center. It would also include the *obligation to assist* the Center and other signatories.

The treaty will also have to deal with the attribution problem, which is not just a matter of nations organizing their citizen hacktivists. The hacktivist problem might be addressed by the provisions in the treaty we have just discussed. Attribution is also a problem because nations route attacks through other countries and sometimes actually initiate them from another nation. The Center could investigate claims by nations that they were not the source of an attack, and it could issue reports to allow the member states to judge if there had been a treaty violation by a particular state. If there had been a clear violation, the states party to the treaty could issue sanctions. The sanctions could range across a spectrum from, at the low end, denying visas or entry to specific individuals, to denying Internet connectivity to an ISP. At the higher end, nations could limit international Internet and telephone traffic flows for a country. The Center could put scanners on the points where traffic from the country came into other nations. Finally, of course, nations could refer the problem to the United Nations and recommend broader economic and other sanctions.

The treaty and the Center would only be concerned with cyber war. It would not become an international regulatory body for the Internet, as some have proposed. Burdening CWLT with that possibility will ensure that it is opposed by many interests in the U.S. and elsewhere. CWLT will not, by itself, stop attacks on civilian targets, but it will raise the price of trying them. The advent of CWLT as an international norm will also send a message to cyber warriors and their government masters that firing off a cyber attack is not the first thing that you do when your neighbor state has made you mad. Engaging in offensive cyber war against another country would become, after CWLT, a major step. Using it against a civilian infrastructure target would be a violation of international law. Nations that signed the CWLT might put in place good internal controls to prevent their own cyber warriors from starting something without proper authorization.

5. CYBERSPACE AT MIDDLE AGE

The fifth element of fighting cyber war is research on more secure network designs. The Internet is now forty, entering midlife, yet it has not changed much from its early days. Yes, bandwidth certainly has grown, as has wireless connectivity, and mobile devices have proliferated. But the underlying design of the Internet, which was done without any serious thought to security, is unaltered. Although many software glitches and security issues were supposed to have disappeared when Microsoft replaced its earlier buggy operating systems with Vista and now Windows 7, problems persist with all of the most ubiquitous software programs.

When I asked the head of network security for AT&T what he would do if someone made him Cyber Czar for a day, he didn't hesitate. "Software." Ed Amoroso sees more security issues in a day than most computer security specialists see in a year. He has written four books on the subject and teaches an engineering course on cyber security. "Software is most of the problem. We have to find a way to write software which has many fewer errors and which is more secure. That's where the government should be funding R&D." Hackers get in where they don't belong, most often because they have obtained "root," or administrator status, through a glitch they have discovered in the software. There are two research priorities created by that phenomenon. We have to do a better job of finding the errors and vulnerabilities in existing software, which is a matter of testing in various ways. But at the same time we need to find a process for writing new applications and operating systems from scratch with close to zero defects.

As much as people fear robots and artificial intelligence (without knowing that there are already a lot of both at work today), it may be worth thinking about using artificial intelligence to write new

code. It would mean coming up with a set of rules for writing secure
and elegant code. The rules would have to be extensive and iterated
with testing. The project would be sufficiently large that it would require government research funding, but it should be possible gradually to develop an artificial intelligence program that could respond
to requests to write software. The artificial code writer could compete with famous software designers, much as IBM's Big Blue played
against human chess masters. Drawing on the open source movement, it could be possible to get the world's experts to contribute to
the process.

The work that was done to create the Internet forty years ago
has been enormously valuable, far more so than the inventors ever
thought then that it would be. Now the funders of the original Internet should fund an attempt to do something better. Today cyber research is fragmented and, according to a presidential advisory
board, cyber security research is dangerously underfunded. Cyberspace also needs a fresh look from designers who are freed to think
of new protocols, new ways of authenticating, and advanced approaches for authorizing access, seamlessly encrypting both traffic
and data at rest.

There are some signs of renewed life at DARPA (the Defense
Advanced Research Projects Agency), which funded much of the
early Internet development. After years of abandoning research on
the public Internet, things have begun to change. In October 2009,
DARPA granted a contract to a consortium including defense contractor Lockheed and router manufacturer Juniper Networks to design a new basic protocol for the Internet. For decades, the Internet
has been breaking traffic up into little digital packets, each with
its own address space, or "header." The header has the basic *to* and
from information. The protocol or format for these packets is named
TCP/IP (Transport Control Protocol/Internet Protocol). For the
gods and founders of the Internet, TCP/IP is as sacred as the Ten

Commandments are to some religious groups. What DARPA is now looking for is something to replace TCP/IP. Shock and horror! The new Military Protocol would allow for authentication of who sent every packet. It would permit prioritization of the packets, depending upon the purpose of the communication. It might even encrypt the content. The Military Protocol would be used initially on the Pentagon's networks, but just think what it could do for the Internet. It could stop most cyber crime, cyber espionage, and much of cyber war. DARPA has no estimated ready date for the Military Protocol, nor any idea about how the conversion process from TCP/IP would occur. Nonetheless, it is just that kind of thinking that could make the Internet secure someday.

We should not throw out what we have until we are sure that the alternative really is better and that the conversion process is feasible. What might that something new look like? In addition to the Internet, cyberspace might consist of many more intranets, but these would be highly heterogeneous, running one of several different protocols. Some of the intranets might have "thin clients," which are not skinny guys looking for a lawyer, but computer terminals that use well-controlled servers or mainframes rather than having an extensive hard drive on every desk. Centralized mainframes (yes, the old mainframe) that, if they failed, would be backed up by redundant hardware at other locations, could manage intranets to prevent security violations and configuration mismanagement at the nodes. The intranets' traffic would run on separate fibers from the public Internet and could be switched by routers that did not touch the public Internet. Data could be scanned for malware and backed up in redundant data farms, some of which would always be disconnected from the network in case of a corrupting system failure. All of these new intranets could use constant scanning technologies to detect and prevent anomalous activity, intrusions, identity theft, malicious software, or unauthorized exporting of data. The intranets could encrypt all data and require that a user prove with two or three reliable methods who he is before he could access the intranet. If the new nets were "packet switched," as the Internet is now, the user's authenticated identity could be embedded in each packet. Most important, these networks could constantly monitor for and prevent connectivity to the Internet.

A lot of people will hate that idea. Many of the Internet's earliest advocates strongly believe that information should be free and freely disseminated, and that essential to that freedom is the right to access information anonymously. The "open Internet" people believe that if you wish to read *The Communist Manifesto*, or research treatments for venereal disease, or document China's human rights violations, or watch porn online, your access to that information will not be free if anyone knows that you are looking at it.

But does that mean that everything should be done on one big, anonymous, open-to-everyone network? That's how Vint Cerf and others see the Internet, and they'll be damned if they're gonna agree to change it. When I worked in the White House, I proposed something I called "Govnet," a private network for the internal working of federal agencies that would deny access to those who could not really prove who they were (maybe with a special fob). Vint Cerf thought that was an awful idea, one that would erode the open Internet, beginning a trend of cutting it up into lots of little networks. Privacy advocates, whose cause I usually support, hated Govnet, too. They thought it would force everyone accessing the public web pages of government agencies to identify themselves. Of course, the public web pages would not have been on Govnet. They would still have been on the public Internet. But in the face of opposition like that, Govnet did not happen. It is probably time that we revisit the Govnet concept now.

In addition to Govnet for critical functions of the federal government, where else might we want such secure networks? For airline operations and air traffic control, railroad operations, medical centers, certain research activities, operations of financial institutions, controlling space flight, and, of course (say it with me), for the power grid. All of these institutions would still need an Internet-facing presence off the intranet, to communicate outside the closed community of the intranet. But there would be no real-time connection between the secure networks and the Internet. Indeed, ideally the protocol, applications, and operating systems would be incompatible.

There would still be a public Internet, of course, and we would all still use it for entertainment, information, buying things, sending e-mail, fighting for human rights, learning about medical problems, looking at pornography, and engaging in cyber crime. But if we worked at a bank, the IRS, or the train company, or (say it loudly) the electric company, we would use one of these new secure, special-purpose intranets when we were at work. Cyber war could still target these intranets, but their diversity, their use of separate routers and fiber, and their highly secured internals would make it very unlikely that they could all be taken down. Vint Cerf and those devoted to one big everybody-goes-everywhere, interconnected web won't like it, but change must come.

6. "IT'S POTUS"

Those were the words our hypothetical White House official heard in chapter 2. Most of the time, those are words you never want to hear, at least when somebody is shoving a phone in your direction in a crisis. The sixth element of our agenda is, however, Presidential involvement. I know that everyone working on a policy issue thinks the President should spend a day a week on his or her pet rock. I don't.

The President should, however, be required to approve personally the emplacement of logic bombs in other nations' networks, as well as approve the creation of trapdoors on a class of politically sensitive targets. Because logic bombs are a demonstration of hostile intent, the President alone should be the one who decides that he or she wants to run the destabilizing risks associated with their placement. The President should be the one to judge the likelihood of the U.S. being in armed conflict with another nation in the foreseeable future, and only if that possibility is high should he or she authorize logic bombs. Key congressional leaders should be informed of such presidential decisions, just as they are for other covert actions. Then, on an annual basis, the President should review the status of all major cyber espionage, cyber war preparation of the battlefield, and cyber defense programs. An annual cyber defense report to the President should spell out the progress made on defending the backbone, securing the DoD networks, and (let me hear you say it) protecting the electric power grid.

In this annual checkup, the President should review what Cyber Command has done: what networks they have penetrated, what options would be available to him in a crisis, and whether there are any modifications needed to his earlier guidance. This review would be similar to the annual covert-action review and the periodic dusting off of the nuclear war plan with the President. Knowing that there is an annual checkup keeps everybody honest. While he is reviewing the cyber war strategy implementation, the President could annually get a report from our proposed Cyber Defense Administration on its progress in securing government agencies, the Tier 1 ISPs, and (all together now) the power grid.

Finally, the President should put reducing Chinese cyber espionage at the top of the diplomatic agenda, and make clear that such behavior amounts to a form of economic warfare.

As I suggested earlier, the President should use the occasion of his annual commencement address at a military service academy, looking out over the cadets or midshipmen and their proud families, to promulgate the Obama Doctrine of Cyber Equivalence, whereby a cyber attack on us will be treated the same as if it were a kinetic attack and that we will respond in the manner we think best, based upon the nature and extent of the provocation. I suggested that he add a proposal for a global system of National Cyber Accountability that would impose on nations the responsibility to deal with cyber criminals and allegedly spontaneous civilian hacktivists, and an Obligation to Assist in stopping and investigating cyber attacks. It would be a sharp contrast to the Bush Doctrine, announced at West Point, that expressed the sentiment that we should feel free to bomb or invade any nation that scares us, even before it does anything to us.

To follow up such a spring speech at an academy, the President should then in September give his annual address at the opening of the United Nations General Assembly session. Looking out from that green granite podium at the leaders or representatives of ninescore countries, he should say that

> The cyber network technology that my nation has given to the world has become a great force for good, advancing global commerce, sharing medical knowledge that has saved millions of lives, exposing human rights violations, shrinking the globe, and, through DNA research, making us more aware that we are all descendants of the same African Eve.
>
> But cyberspace has also been abused, as a playground for criminals, a place where billions of dollars are annually siphoned off to support cartels' illicit activities. And it has already been used by some as a battlespace. Because cyber weapons are so easily activated and the identity of an attacker can sometimes be kept secret, because cyber weapons can strike thousands of targets and inflict extensive disruption and damage in seconds, they are potentially a

new source of instability in a crisis, and could become a new threat to peace.

Make no mistake about it, my nation will defend itself and its allies in cyberspace as elsewhere. We will consider an attack upon us through cyberspace as equivalent to any other attack and will respond in a manner we believe appropriate based on the provocation. But we are willing, as well, to pledge in a treaty that we will not be the first in a conflict to use cyber weapons to attack civilian targets. We would pledge that and more, to aid in the creation of a new international Cyber Risk Reduction Center, and undertake obligations to assist other nations being victimized by attacks originating in cyberspace.

Cyber weapons are not, as some have claimed, simply the next stage in the evolution of making war less lethal. If they are not properly controlled, they may result in small disagreements spiraling out of control and leading to wider war. And our goal as signers of the United Nations Charter is, as pledged in San Francisco well over half a century ago, "to save succeeding generations from the scourge of war." I ask you to join me in taking a step back from the edge of what could be a new battlespace, and take steps not to fight in cyberspace, but to fight against cyber war.

It could be a beautiful speech, and it could make us safer.

Glossary

A Guide to the Cyber Warrior's Acronyms and Phrases

Authentication: Procedures that attempt to verify that a network user is who he or she claims to be. A simple authentication procedure is a password, but software can be used to discover passwords. "Two-factor" authentication is the use of a password and something else, such as a fingerprint or a series of digits generated by a fob, a small handheld device.

Backbone: The Internet backbone consists in the coast-to-coast trunk cables of fiber optics, referred to as "big pipes," run by the Tier 1 ISPs.

Border Gateway Protocol (BGP): The software system by which an ISP informs other ISPs who its clients are so that messages intended for the client can be routed or switched to the appropriate ISP. Sometimes an ISP may have other ISPs as clients. Thus, for example, AT&T may list on its BGP table an Australian ISP. If a packet originates on, for example, Verizon, and Verizon does not connect to the Australian network, a Verizon router at a telecom hotel (*see below*) would look at a BGP table to see who does have such a connection and would, in this example, route the packet to AT&T for onward routing to the Australian network. BGP tables are not highly secure and can be spoofed, leading to the misrouting of data.

Botnet: A network of computers that have been forced to operate on the commands of an unauthorized remote user, usually without the knowledge of their owners or operators. This network of "robot" computers is then used to commit attacks on other systems. A botnet usually has one or more controller computers, which are being directly employed by the operator behind the botnet to give orders to the secretly controlled devices. The computers on botnets are frequently referred to as "zombies." Botnets are used, among other purposes, to conduct floods of messages (*see* DDOS).

Buffer Overflow: A frequent error in computer code writing that allows for unauthorized user access to a network. The error is a failure to limit the number of characters that can be entered by a nontrusted user, thus allowing such a user to enter instructions to the software system. For example, a visitor to a webpage may go to a section of the page where he should only be able to enter his address and instead enters instructions that allow him to gain the same access as the network's administrator.

Civilian Infrastructure: Those national systems that make it possible for the nation's economy to operate, such as electric power, pipelines, railroads, aviation, telephony, and banking. In the U.S., these separate verticals usually consist of nongovernmental entities, privately held or publicly traded corporations that own and/or operate the systems.

Crisis Instability: In a period of rising tensions or hostilities between nations, there may be preconditions or actions taken by one side that cause the other nation to believe it is in its best interest to take further aggressive action. Crisis instability is that condition that may lead to decisions to escalate military actions.

Cyber Boundary: The cyber/kinetic boundary is the decision point when a commander must decide whether and how to move from a purely cyber war to one involving conventional forces, or kinetic weapons. Crossing the boundary is an escalatory step that may lead to the war spiraling out of control.

DARPA (also seen as ARPA): The Defense Advanced Research Projects Agency is a component of the U.S. Defense Department charged with funding innovative research to meet the needs of the U.S. military. DARPA funded the initial research that created the Internet. In 1969 ARPANET became the first packet-switched network connecting four universities.

Deep-Packet Inspection: A procedure that scans the packets of data that make up an e-mail, webpage, or other Internet traffic. Normally only the "header" of a packet is scanned, the top part that gives the *to* and *from* information. A deep inspection would scan the digital pattern in the content but would not convert that content into text. The inspection looks only for digital patterns that are identical or highly similar to known malware or hacking tools.

Distributed Denial of Service (DDOS): A basic cyber war technique often used by criminals and other nonstate actors in which an Internet site, a server, or a router is flooded with more requests for data than the site can respond to or process. The result of such a flood is that legitimate traffic cannot access the site and the site is in effect shut down. Botnets are used to conduct such attacks, thus "distributing" the attack over thousands of originating computers acting in unison.

Domain Name System (DNS): A hierarchy of computers that converts words used as Internet addresses (as in www.google.com) into the numerical addresses that the networks actually use for routing message traffic (as in 192.60.521.7294). At the lowest rung of the hierarchy a DNS server may know only the routing information within a company; at a higher level a computer might know routing information for within a "domain," such as the dot-net (.net) set of addresses. The highest-level DNS computers may contain the routing information for a national domain, such as dot-de (.de) for Germany— the "de" standing, of course, for "Deutschland." DNS computers are vulnerable to floods of demands (*see* DDOS) and to unauthorized changes in routing information, or "spoofing," in which a user is sent to a fraudulent look-alike version of the intended webpage.

Edge: That place on the Internet where local traffic connects to a larger, nationally connected fiber-optic cable. An edge router directs locally originating traffic onto the national network.

Encryption: The scrambling of information so that it is unreadable to those who do not have the code to unscramble it. Encrypting traffic (or "data at rest") prevents those who intercept it or steal it from being able to read it.

Equivalence: The Cyber Equivalence Doctrine is a policy under which a cyber war attack will be treated like any other attack, including a kinetic strike, and will be responded to in a manner of the attacked nation's own choosing, based upon the extent of the damage done and other relevant factors.

Escalation Dominance: When one party to a conflict responds to an attack or provocation by significantly expanding the scope or level of the conflict and at the same time communicates that if its demands (such as war termination) are not met it can and will go even further, this is referred to as "escalation dominance." The expansion of the hostilities is meant to demonstrate seriousness of intent and strength of capability, as well as a refusal to tolerate a prolonged low-level conflict. It is similar to the poker move of significantly raising the stakes and bringing the contest to an end-game phase in the hopes of convincing an opponent to back down.

Espionage: Intelligence activities designed to collect information, access to which another nation (or other actor) is attempting to deny. Cyber espionage is the unauthorized entry by a nation-state onto the networks, computers, or databases of another nation for purposes of copying and exfiltrating sensitive information.

Hacker: Originally, a skilled user of software or hardware who can adapt systems to do things other than their intended or original use. In common parlance, however, the term has been used to denote someone who uses skills to gain access to a computer or network without authorization. As a verb, "to hack" means to break into a system.

Internet: The global interconnected network of networks intended for general access for the transmission of e-mails, the sharing of information on webpages, and so on. Networks may use the same software and transmission protocols, but not be part of the Internet if they are designed to be closed off from the global interconnected system. Such closed networks are referred to as "intranets." Often there are controlled connections between intranets and the Internet. Sometimes there are unintentional connections.

Internet Service Provider (ISP): A corporation (or government agency) that provides the wired or wireless connectivity from a user's home, office, or mobile computer to the Internet. In the U.S. there are numerous small, regional ISPs and a handful of national ISPs. Often ISPs are also telephone companies or cable television providers.

JWICS: The Joint Worldwide Intelligence Communication System is the Defense Department's global intranet for transmitting data that it has classified Top Secret/SCI (Specially Compartmented Information). TS/SCI information is derived from intelligence collection systems such as satellites (*see* NIPRNET *and* SIPRNET).

Latency: The extent to which a data packet is slowed from moving as quickly as possible on a network or path. Latency is measured in seconds or parts of seconds. The fastest, unimpeded speed is referred to as "line rate." The size of a fiber-optic cable and the processing speed of routers along a network determine the line rate for that cable and/or router.

Launch on Warning: A strategy component that dictates that a nation will initiate conflict—in this case, a cyber war—when intelligence indicators suggest that an opponent has or is about to commence hostile activities.

Logic Bomb: A software application or series of instructions that cause a system or network to shut down and/or to erase all data or software on the network.

Malware: Malicious software that causes computers or networks to do things that their owners or users would not want done. Examples of malware include logic bombs, worms, viruses, packet sniffers, and keystroke loggers.

National Accountability: The concept that a national government will be held responsible for cyber attacks originating inside its physical boundaries. Also called the Arsonist in the Basement Theory ("If you are harboring an arsonist in your house and he is going out from your house and burning down others, you are just as responsible as he is").

National Cyber Strength: A net assessment of a nation's ability to fight cyber war, the national cyber strength takes into account three factors: offensive cyber capability, the nation's dependence upon cyber networks, and the ability of the nation to control and defend its cyberspace through such measures as cutting off traffic from outside the country.

NIPRNET: Non-classified Internet Protocol Router Network is the Defense Department's global intranet for information that is not classified. NIPRNET connects with the Internet at a limited number of portals. These are two other Defense Department intranets, SIPRNET and JWICS.

No First Use: In arms control, the concept that a nation will not employ a certain kind of weaponry until and unless it has been used on it. Implicit in the concept is that a nation will only use a certain kind of weapon on those that have already used it, and that the use of the weapon would be an in-kind retaliation.

NSA: The National Security Agency is a U.S. intelligence agency that is also a component of the Defense Department. NSA is the lead U.S. agency for collecting information through electronic means. It is headquartered at Fort Meade, Maryland, and is frequently referred to simply as "The Fort."

Obligation to Assist: The proposal that each nation in a cyber war agreement would take on a requirement to help other nations and/or the appropriate international body in investigating and stopping cyber attacks originating from within its own physical boundaries.

Out of Band: Communications, frequently about the management of a network, that use a different channel or method of communicating than the network being managed.

Server: A computer usually accessed by many others, in order to interact with information stored on it, such as web pages or e-mails. Typically, servers are meant to operate without constant human monitoring. Routers, which direct the movement of Internet traffic, are a type of server.

SIPRNET: Secret Internet Protocol Router Network is the Defense Department's global intranet for transmitting confidential and secret-level information. The Defense Department classifies information into five catergories: unclassified, confidential, secret, top secret, top secret/SCI (specially compartmented information). The SIPRNET is supposed to be air-gapped from, i.e., not physically touching, the unclassified NIPRNET and the Internet.

Supervisory Control and Data Acquisition System (SCADA):
Software for networks of devices that control the operation of a system of machines such as valves, pumps, generators, transformers, and
robotic arms. SCADA software collects information about the condition of and activities on a system. SCADA software sends instructions to devices, often to do physical movements. Instructions sent to
devices on SCADA networks are sometimes sent over the Internet or
broadcast via radio waves. Instructions are not encrypted. When the
devices receive orders, they do not validate who sent the instructions.

TCP/IP: Transmission Control Protocol/Internet Protocol. The
format used to divide information such as e-mails into digital "packets," each with its own *to* and *from* data so that the packet can be
routed on the Internet.

Telecom Hotels: Buildings that house large numbers of network
routers, often places where major networks connect to each other.
Internet and other cyber traffic, including voice telephony, are
switched in such a facility. Large telecom hotels are sometimes called
gigapops (points of presence). Early Internet switching centers were
called Metropolitan Area Exchanges (MAEs); two examples are
MAE East in Tysons Corner, Virginia, and MAE West in San Jose,
California.

Tier 1: The five Internet service providers (ISPs) in the U.S. that
own and operate the large, national network of fiber-optic cables on
which Internet and other cyberspace traffic runs to the major cities.
Smaller or regional ISPs use a Tier 1 to connect to Internet addresses
that are on their own network.

Trapdoor: Unauthorized software maliciously added to a program
to allow unauthorized entry into a network or into the software

program. Often after an initial entry, a cyber criminal or cyber warrior leaves behind a trapdoor to permit future access to be faster and easier. Also referred to as a Trojan, or Trojan horse, after a ruse supposedly employed by Bronze Age Greek warriors to leave behind at Troy a commando team hidden inside a statue of a horse.